Current Topics in Microbiology
and Immunology

111

Editors

M. Cooper, Birmingham/Alabama · W. Goebel, Würzburg
P.H. Hofschneider, Martinsried · H. Koprowski, Philadelphia
F. Melchers, Basel · M. Oldstone, La Jolla/California
R. Rott, Gießen · H.G. Schweiger, Ladenburg/Heidelberg
P.K. Vogt, Los Angeles · R. Zinkernagel, Zürich

The Molecular Biology of Adenoviruses 3

30 Years of Adenovirus Research 1953–1983

Edited by Walter Doerfler

With 26 Figures

Springer-Verlag
Berlin Heidelberg New York Tokyo 1984

Professor Dr. WALTER DOERFLER
Institut für Genetik
der Universität zu Köln
Weyertal 121
D-5000 Köln 41, FRG

ISBN-13:978-3-642-69551-3   e-ISBN-13:978-3-642-69549-0
DOI: 10.1007/978-3-642-69549-0

© by Springer-Verlag Berlin Heidelberg 1984
Softcover reprint of the hardcover 1st edition 1984

Library of Congress Catalog Card Number 15-12910

2123/3130-543210

Table of Contents

Indexed in Current Contents

List of Contributors

ESCHE, H., Institute of Genetics, University of Cologne, Weyertal 121, D-5000 Köln 41

FROLOVA, E.I., Institute of Molecular Biology, The Academy of Sciences of the USSR, Moscow, USSR

FÜTTERER, J., Institut für Biochemie der Universität München, Karlstr. 23, D-8000 München 2

REUTHER, M., Institute of Genetics, University of Cologne, Weyertal 121, D-5000 Köln 41

SCHUGHART, K., Institute of Genetics, University of Cologne, Weyertal 121, D-5000 Köln 41

SHENK, T., Department of Microbiology, Health Sciences Center, State University of New York, Stony Brook, NY 11794, USA

WILLIAMS, J., Department of Biological Sciences, Carnegie-Mellon University, Pittsburgh, PA 15213, USA

WINNACKER, E.-L., Institut für Biochemie der Universität München, Karlstr. 23, D-8000 München 2

ZALMANZON, E.S., Institute of Molecular Biology, The Academy of Sciences of the USSR, Moscow, USSR

Genetic Analysis of Adenoviruses

T. Shenk[1] and J. Williams[2]

1 Introduction

Adenovirus mutants have provided key insights into the functional roles of viral gene products. They have facilitated the dissection of numerous viral processes, including transcription, translation, DNA replication, assembly, and transformation. Additional variants will allow these processes to be studied in ever greater detail.

The aim of this review is to describe some of the wide variety of adenovirus mutants that have been isolated and to discuss the physiological consequences of their alterations. We have chosen to limit the scope of our discussion to the two most studied human serotypes, adenoviruses type 2 and type 5 (Ad2 and Ad5). Table 1 lists the majority of mutants described and provides key references. Additional aspects of adenovirus genetics are discussed in a recent review by Young et al. (1983).

---

1 Department of Microbiology, Health Sciences Center, State University of New York, Stony Brook, NY 11794, USA
2 Department of Biological Sciences, Carnegie-Mellon University, Pittsburgh, PA 15213, USA

Current Topics in Microbiology and Immunology, Vol. 111
© Springer-Verlag Berlin·Heidelberg 1984

Table 1. Selected Ad2 and Ad5 mutants

Location	Mutant	References
E1A	H5*in*340-2,A5,B1	HEARING and SHENK 1983a
	H5*dl*311,312	JONES and SHENK 1979a, b; SHENK et al. 1979; ESCHE et al. 1980; ROSS et al. 1980a; NEVINS 1981; GAYNOR and BERK 1983
	H2/5*dl*1504	OSBORNE et al. 1982
	H5*dl*347,348	WINBERG and SHENK, unpublished
	H2/5*pm*975	MONTELL et al. 1982
	H5*hr*440	SOLNICK 1981; SOLNICK and ANDERSON 1982
	H5*dl*101,105	BABISS et al. 1984
	H5*in*500	CARLOCK and JONES 1981
	H5*hr*1	HARRISON et al. 1977; BERK et al. 1979; ESCHE et al. 1980; GALOS et al. 1980; ROSS et al. 1980a; RICCIARDI et al. 1981; KATZE et al. 1981; HO et al. 1982; BABISS et al. 1983a
	H2*pm*1610	MONTELL et al. 1983
E1B	H5*dl*313	JONES and SHENK 1979a; SHENK et al. 1979; ESCHE et al. 1980; ROSS et al. 1980b; COLBY and SHENK 1981; SHIROKI et al. 1981; LAI FATT and MAK 1982; MAK and MAK 1983
	H2*lp*3	CHINNADURAI 1983
	H5*dl*337,338	PILDER, LOGAN, and SHENK, unpublished
	H5*hr*6,7	HARRISON et al. 1977; GRAHAM et al. 1978; LASSAM et al. 1978, 1979; GALOS et al. 1980; ROSS et al. 1980
	H5*hr^{cs}*13	HO et al. 1982
E2A	H5*ts*107,125	ENSINGER and GINSBERG 1972; GINSBERG et al. 1974; VAN DER VLIET et al. 1975; CARTER and GINSBERG 1976; MAYER and GINSBERG 1977; CARTER and BLANTON 1978b; HORWITZ 1978; KAPLAN et al. 1979; NEVINS and WINKLER 1980; BABICH and NEVINS 1981; KRUIJER et al. 1981.
	H2*ts*23	KRUIJER et al. 1982; WILLIAMS and FRASER unpublished work
	H5r(*ts* 125)13,20	NICOLAS et al. 1981
	H5r(*ts* 107)202	CARTER et al. 1982; NICOLAS et al. 1982; KRUIJER et al. 1983
	H5*hr*404	KLESSIG and GRODZICKER 1979; KRUIJER et al. 1981; KLESSIG and QUINLAN 1982
E2B	H5*ts*36,69,149	RUSSELL et al. 1972; WILKIE et al. 1973; GINSBERG et al. 1974; WILLIAMS et al. 1974; CARTER and GINSBERG 1976; GALOS et al. 1979; STILLMAN et al. 1982
E3	H5*dl*324,*Δ*E3	THIMMAPPAYA et al. 1982; BERKNER and SHARP 1984
E4	H2*dl*807,808	CHALLBERG and KETNER 1981
	H5*dl*350-359	HALBERT, CUTT and SHENK, unpublished work
L1	H5*ts*58	EDVARDSSON et al. 1978
L1 or L2	H2*ts*112	D'HALLUIN et al. 1978, 1982; MARTIN et al. 1978

Table 1 (continued)

Location	Mutant	References
L3	H5ts2,40	WILLIAMS et al. 1971; RUSSELL et al. 1972, 1974
	H2ts1	BEGIN and WEBER 1975; WEBER 1976; YEH-KAI et al. 1983
L4 or E2A	H5ts19	WILLIAMS et al. 1971; USTACELEBI and WILLIAMS 1972; EDVARDSSON et al. 1978; TARODI et al. 1979
L4	NDlts4	SAMBROOK et al. 1975; GRODZICKER et al. 1977; OOSTEROM-DRAGON and GINSBERG 1981
	H5ts1,18,24,115	WILLIAMS et al. 1971; ENSINGER and GINSBERG 1972; RUSSELL et al. 1972; WILLIAMS et al. 1974; EDVARDSSON et al. 1978; TARODI et al. 1979; OOSTEROM-DRAGON and GINSBERG 1981
	H2ts48,118	HASSELL and WEBER 1978; MARTIN et al. 1978; CARSTENS et al. 1979; D'HALLIUM et al. 1982
L5	H5ts5,9,22,142	WILLIAMS et al. 1971; RUSSELL et al. 1972; GINSBERG et al. 1974; EDVARDSSON et al. 1978; CHEE-SHEUNG and GINSBERG 1982
VA	H5dl309	THIMMAPPAYA et al. 1979
	H5dl328,331	THIMMAPPAYA et al. 1982
	H5dl706,707,710, 721,sub720	BHAT and THIMMAPPAYA, unpublished

2 Production of Viral Mutants

A wide variety of procedures have been utilized to generate adenovirus mutants. Initially, chemical and physical mutagens were used to generate variants. This approach has the advantage that sophisticated understanding of the physical organization of a viral chromosome is not necessary, and it rapidly provides information about essential viral functions. The use of mutagens for production of adenovirus mutants has been reviewed in detail previously (GINSBERG and YOUNG 1977) and need not be discussed further here.

More recently, mutagenesis schemes have been developed that are based on in vitro manipulation of the viral genome. These procedures generally require a reasonably detailed understanding of the viral chromosome's architecture. Many deletion and substitution mutants have been isolated simply by selecting variants that lack restriction endonuclease cleavage sites (JONES and SHENK 1978, 1979a; MATHEWS and GRODZICKER 1981; RAJAGOPALAN and CHINNADURAI 1981). As efficient methods became available for reconstructing viral DNA segments into intact chromosomes (CHINNADURAI et al. 1979; STOW 1981) it became feasible to construct a wide variety of precisely positioned mutations (e.g. SOLNICK 1981; STOW 1981; HO et al. 1982; MONTELL et al. 1982; OSBORNE et al. 1982; THIMMAPPAYA et al. 1982; HEARING and SHENK 1983a, b). In fact, it is now possible to manipulate a cloned segment derived from any location on the viral genome and ultimately rebuild the altered region into an intact chromosome for study. This, of course, makes it possible to utilize any of the available mutagenesis procedures (reviewed in SHORTLE et al. 1981).

A number of procedures have been utilized for propagating viral mutants. In the simplest case, the mutant is viable and may or may not grow as well as the wild-type virus (e.g., KELLY and LEWIS 1973; YOUNG and WILLIAMS 1975; JONES and SHENK 1978; THIMMAPPAYA et al. 1982; CHINNADURAI 1983). Conditionally defective viruses have been produced and propagated in large numbers. These include temperature-sensitive (GINSBERG and YOUNG 1977) and host-range mutants (e.g., HARRISON et al. 1977; JONES and SHENK 1979a). Only one instance has been reported in which a nonconditionally defective mutant was propagated and studied. CHALLBERG and KETNER (1981), using a temperature-sensitive helper virus, propagated variants carrying large deletions. Clearly, additional methods for propagating defective viral mutants must be developed, and this topic is addressed near the end of this review.

3 Genetic and Physical Maps

Meaningful functional analysis of a mutant demands that the mutation be mapped as precisely as possible on the genome. As a first step, recombination frequencies can be generated from a series of two-factor crosses between mutants under nonselective conditions, and a genetic map giving relative distances between mutants can be constructed (e.g., WILLIAMS et al. 1974). This map can then be oriented with respect to the conventional adenovirus genomic map, either by the method of intertypic recombination (GRODZICKER et al. 1974; WILLIAMS et al. 1975; SAMBROOK et al. 1975) or by marker rescue (ARRAND 1978; FROST and WILLIAMS 1978). The first method depends upon the presence of distinctive restriction enzyme cleavage sites on the genomes of two serotypes capable of recombination. The method can be quite precise, but is limited by availability of differential cleavage sites. Marker rescue is, in practice, the simple co-transfection of an appropriate cell line with intact genomic DNA from the mutant and pure restriction fragments representing segments of the wild-type genome. Should a single mutation be responsible for the mutant phenotype, then one subgenomic fragment will rescue the ts^+ allele and generate wild-type progeny. Sets of overlapping fragments generated by different enzymes can be used to obtain excellent resolution in some cases. Further, the method is capable of revealing possible second-site mutations elsewhere on the genome should they possess phenotypic identity.

As an alternative to marker rescue, the technique of overlap recombination can be used. In this case, two terminal DNA fragments with overlapping sequences, which are individually noninfectious, produce infectious progeny when co-transfected into appropriate cells (CHINNADURAI et al. 1979; Ho et al. 1982). VOLKERT and YOUNG (1983) have demonstrated that distances between mutations within the overlap region can be determined, and that the mutations can be ordered. Segregation of closely spaced pairs of mutations can be achieved.

In addition to these methods, the deletion mapping method of BENZER (1961) has been applied to adenovirus (GALOS et al. 1980). Precisely mapped deletion mutants are crossed with point mutants under permissive conditions. Wild-type recombinants occur only if the mutation lies outside the deletion limits, and no recombinants form if the mutation lies within these limits. Deletions made in cloned adenovirus fragments can be used in marker rescue analyses to verify and refine limits determined by conventional marker rescue or other methods (B. KARGER, B. MILLER, and J. WILLIAMS, unpublished results). There is, in general, excellent agreement between genetic and physical maps (cf. WILLIAMS et al. 1974; FROST and WILLIAMS 1978; see also GINSBERG and YOUNG 1977).

The various mapping methods mentioned above assign the mutated site to a small segment of the genome, and may allow unambiguous assignment to the reading frame for a particular protein. Overlapping or very close reading frames abound in adenovirus, and it is necessary in most cases to determine the site of the mutation exactly by nucleotide sequence analysis (MAXAM and GILBERT 1977; SANGER et al. 1977). This has now been carried out for a number of mutants that were generated by random mutagenesis (KRUIJER et al. 1981; RICCIARDI et al. 1981; YEH-KAI et al. 1983; cf. figures in this review). More detailed discussion of mapping methods has been given by GINSBERG and YOUNG (1977) and YOUNG et al. (1983).

4 Physiological Consequences of Mutations

4.1 Region E1A Lytic Functions

The E1A transcription unit lies at the extreme left end of the viral chromosome and is apparently the first to be expressed subsequent to infection. It encodes three mRNAs, which have common 5′ and 3′ ends but differ in their splicing patterns (Fig. 1). The three mRNAs (designated 13S, 12S, and 9S) contain coding regions which would specify 31.9K-, 26.5K-, and 13.3K-dalton polypeptides, respectively (MAAT and VAN ORMONDT 1979; GINGERAS et al. 1982). The 12S and 13S mRNAs encode a family of four to six polypeptides in vivo with apparent molecular weights of 40K–60K (e.g., SMART et al. 1981; YEE et al. 1983; FELDMAN and NEVINS 1983), which are localized in both nucleus and cytoplasm of the infected cell. The reasons for the heterogeneity and slower-than-predicted electrophoretic mobility are not yet clear. The 9S mRNA is expressed only later after infection, and its polypeptide product has not been identified within infected cells.

The majority of mutants carrying lesions in the E1A region have been isolated as host-range mutants. These variants grow well in a human cell line called 293 cells but very poorly in other human cell lines. The 293 cells are human embryonic kidney cells which contain and express regions E1A and E1B from Ad5 (GRAHAM et al. 1977; AIELLO et al. 1979). As a result, they serve as a permissive cell for propagation of viruses defective in E1 function. Initially, two groups of Ad5 host-range mutants were isolated. The first had

EIA TRANSCRIPTION UNIT

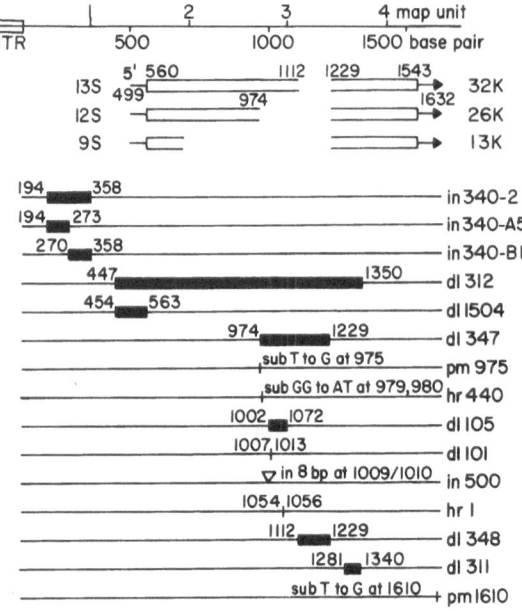

Fig. 1. Physical map of E1A mRNAs, coding regions, and mutations which alter E1A expression. The *top* of the figure positions the map in terms of map units and nucleotide sequence position relative to the left end of the viral chromosome (MAAT and VAN ORMONDT 1979; GINGERAS et al. 1982). *ITR*, inverted terminal repeat sequence. mRNAs are indicated by *lines*, intervening sequences by *spaces*, and polypeptide coding regions by *open boxes*. Deletion mutations are represented by *closed boxes*, insertion mutations by *triangles*, and small alterations (substitutions, small deletions) by *bars*. The first and last nucleotides which are present bracketing alterations are designated by their *nucleotide sequence numbers*, except in the case of substitutions where the altered nucleotides are designated. References for the mutations are provided in Table 1

point mutations (HARRISON et al. 1977; GRAHAM et al. 1978), which were produced by treatment with nitrous acid or ultraviolet light, and their isolation was based on their host-range phenotype. The second group had deletion and substitution mutations (JONES and SHENK 1979a; SHENK et al. 1979). These mutants were selected as lacking the *Xba*I cleavage site at 3.8 map units on the Ad5 chromosome. The prototype of the first group, *hr*1, and that of the second group, *dl*312, are shown in diagram form in Fig. 1.

Analysis of the *hr*1 and *dl*312 growth defect led to the conclusion that region E1A encodes a regulatory function required for expression of non-E1A transcription units. This was first shown by analysis of steady-state levels of mRNA in virus-infected HeLa cells (BERK et al. 1979; JONES and SHENK 1979b). Viruses carrying mutations within E1A produced dramatically reduces levels of all remaining early mRNAs, and as predicted the polypeptides produced by these early mRNAs were also reduced (ROSS et al. 1980b). It is not entirely clear whether the E1A product functions at the level of transcriptional initiation. NEVINS (1981) measured transcription rates and concluded the rate of synthesis of virus-specific mRNAs was reduced in *hr*1 and *dl*312-infected cells. KATZE et al. (1981) found no difference in the rate of nuclear RNA synthesis when comparing hr1 and wild-type-infected cells. The reason for the disagreement is not apparent. It is clear, however, that the E1A product is not absolutely essential for expression of additional viral transcription units. Normal virus yields are produced when HeLa cells are infected at very high multiplicity (e.g., 800 plaque-forming units per cell, SHENK et al. 1979), and additional transcription units turn on slowly after long periods of time, even when cells are infected at relatively low multiplicities (NEVINS 1981; GAYNOR and BERK 1983).

Various early transcription units exhibit differential sensitivity to regulation by the E1A gene product. Wild-type virus synthesizes E1A, E1B, and E4 mRNAs at normal rates, but E2 and E3 species at reduced rates when infections are performed in the presence of anisomycin, an inhibitor of protein synthesis (NEVINS 1981; PERSSON et al. 1981; SHAW and ZIFF 1982). Further, hr440 (SOLNICK 1981; SOLNICK and ANDERSON 1982), which carries several base-pair changes in E1A, and dl1551 (OSBORNE et al. 1982), whose lesion reduced the level of E1A expression, both produce near-normal levels of E1B and E4 mRNAs but reduced amounts of E2 and E3 species. Taken together, the mutant and drug experiments suggest that E1B and E4 are less dependent on E1A function than are E2 and E3.

The regulatory gene product is encoded by the 13S mRNA. Several facts support this assertion. First, hr1 and in500 (CARLOCK and JONES 1981), which exhibit the regulatory defect, carry lesions within the region unique to the 13S mRNA (Fig. 1). Second, pm975 (MONTELL et al. 1982), which contains a single base-pair change within the 12S mRNA splice donor site, cannot produce 12S mRNA and does not display the regulatory defect. Third, a virus that can only produce the 12S mRNA (dl347, constructed using a 12S cDNA, WINBERG and SHENK, unpublished work) fails to turn on additional transcription units. As yet, there is no formal proof that the regulatory gene product is a polypeptide as opposed to the 13S mRNA itself. However, it seems fair to assume that the active agent is a polypeptide since a single base-pair deletion (hr1), which should have a relatively minor effect on the mRNA structure, can destroy the E1A function.

What is the mechanism of the E1A regulatory effect? So far, the literature on this point is contradictory. As mentioned above, agreement has not yet been reached as to whether the effect occurs at the level of transcription initiation or after the primary transcript has been produced. Further, NEVINS (1981) reported that HeLa cells pretreated with cycloheximide and then infected with dl312 synthesized E4-specific mRNAs at normal rates, but not E1B, E2 or E3 species. Normally, of course, dl312 does not direct the synthesis of mRNAs in HeLa cells. KATZE et al. (1981) did a similar experiment, pretreating cells with anisomycin before infection with hr1, and found the rate of transcription unchanged while the steady-state cytoplasmic levels of viral mRNAs increased. Thus, while both groups reported an effect due to treatment of cells with an inhibitor of protein synthesis prior to infection, NEVINS (1981) concluded that it occurred at the level of transcription while KATZE et al. (1981) placed the effect posttranscriptionally. As a result, one group postulated that the E1A regulatory product normally antagonizes a cellular transcriptional repressor while the other suggested that the product acted to inhibit a cellular function which destabilizes viral mRNAs. More recently, GAYNOR and BERK (1983) repeated this experiment and found no changes in the cytoplasmic level of viral mRNAs, contradicting both the previous results. The basis for the contradictory results is not yet clear. GAYNOR and BERK (1983) have suggested that the E1A gene product might facilitate a *cis*-acting modification of the viral template subsequent to infection. This change might either involve removal of virion polypeptides which inhibit transcription or result from the assembly of viral DNA into stable transcription complexes.

Whatever the mechanism of E1A function, it is likely that a variety of nonadenovirus genes is regulated by a similar mechanism. Segments of the adenovirus genome containing only the E1A transcription unit can immortalize primary rat cells (HOUWELING et al. 1980), and E1A plays a direct role in transformation (discussed below). Both these effects suggest that expression of normal cellular genes can be altered by the E1A function. Further, infection of HeLa cells with wild-type Ad5, but not *dl*312, induces synthesis of a cellular 70K heat shock protein (NEVINS 1982), again suggesting that certain cellular genes can be modulated by the E1A function. Finally, pseudorabies virus can complement the *dl*312 transcriptional defect (FELDMAN et al. 1982), suggesting that herpesviruses either express a gene product with an E1A-like function or switch on a cellular gene with such function.

The product of the 12S mRNA is nonessential for viral growth in cultured cells, since both *pm*975 (MONTELL et al. 1982) and *dl*348 (WINBERG and SHENK, unpublished work) make only the 13S species and grow normally. The 9S mRNA is also nonessential, since *dl*312 grows well on 293 cells which do not synthesize detectable levels of this RNA.

4.2 Region E1B Lytic Functions

The E1B transcription unit lies directly to the 3′ side of the E1A unit (Fig. 2). It encodes two mRNAs (22S and 13S) early after infection. The 22S mRNA can code both a 21K and a 55K polypeptide, while the 13S mRNA codes only the 21K species (BOS et al. 1981). The 21K coding region begins just 13 nucleotides downstream of the mRNA 5′ end. The 55K coding region overlaps the C-terminal 40% of the 21K region in a second reading frame. The 55K polypeptide is located in the nucleus and cytoplasm (YEE et al. 1983), while the 21K moiety is found in the plasma membrane of infected cells (PERSSON et al. 1982). The 55K polypeptide associates with both the cellular p53 antigen that was originally described in a complex with SV40 large T antigen (SARNOW et al. 1982b) and with the E4 34K polypeptide (SARNOW et al., to be published). The synthesis of a third mRNA (9S) is directed by a control region within E1B, which becomes active late after infection. This small, unspliced mRNA encodes a 14K structural polypeptide (IX).

As was the case for E1A, the first E1B mutants were isolated as host-range variants using 293 cells as the permissive cell. *hr*6 and *hr*7 are prototypes of this group (HARRISON et al. 1977; GRAHAM et al. 1978). These two mutants

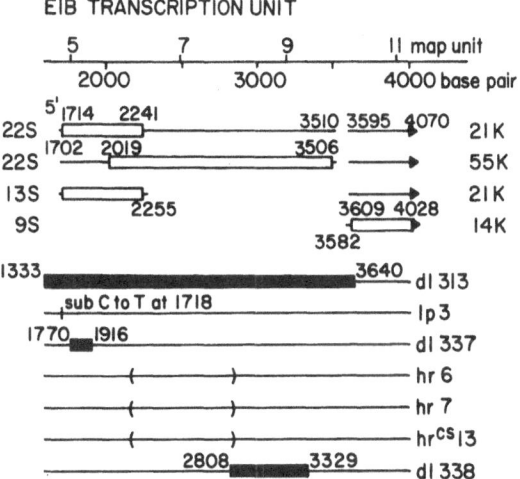

EIB TRANSCRIPTION UNIT

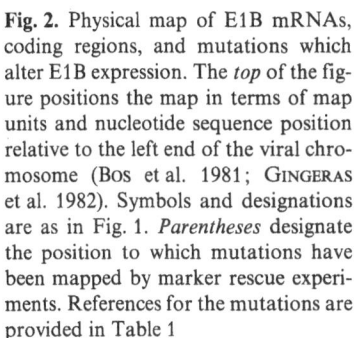

Fig. 2. Physical map of E1B mRNAs, coding regions, and mutations which alter E1B expression. The *top* of the figure positions the map in terms of map units and nucleotide sequence position relative to the left end of the viral chromosome (Bos et al. 1981; GINGERAS et al. 1982). Symbols and designations are as in Fig. 1. *Parentheses* designate the position to which mutations have been mapped by marker rescue experiments. References for the mutations are provided in Table 1

have been localized by recombinational mapping with deletion mutants and by marker rescue to the interval between 6.1 and 8 map units (Fig. 2) (GALOS et al. 1980). These mutants fail to produce the 55K polypeptide (LASSAM et al. 1978; Ross et al. 1980b). So far, it is not clear whether the *hr*6 and *hr*7 lesions affect both the 21K and 55K or only the 55K polypeptide. *hr*cs13 maps to the same interval (Ho et al. 1982), and both *hr*cs13 and *hr*6 are cold-sensitive for growth in HeLa cells. *dl*313 was isolated as lacking the 3.8 map unit *Xba*I cleavage site (JONES and SHENK 1979a). Its large deletion removes the C terminus of E1A proteins, the entire E1B 21K and 55K coding regions, and the N terminus of protein IX. This mutant complements *dl*312, suggesting that it expresses functional E1A products, and its phenotype should, in the main, result from its inability to produce E1B products. *lp*3 is a member of a group of large-plaque variants isolated after chemical mutagenesis of viral particles, and its lesion lies exclusively within the 21K coding region (CHINNADURAI 1983). Finally, *dl*337 and *dl*338 carry deletions localized to the 21K and 55K coding regions, respectively (S. PILDER, J. LOGAN, and T. SHENK, unpublished work). These mutants are shown in diagram form in Fig. 2. Mutants carrying lesions in the E1B 21K polypeptide, *lp*3 and *dl*337, grow well on HeLa cells – in fact they make larger-than-normal plaques on these cells. The large-plaque phenotype suggests they may be related to the Ad12 *cyt* mutants (TAKEMORI et al. 1968). This makes sense, because LAI FATT and MAK (1982) have localized the *cyt* mutations to the region deleted in *dl*313. These authors also showed that extensive degradation of both host and viral DNA occurs in *dl*313-infected cells, as is the case for Ad12 *cyt* mutants. As yet, *lp*3 and *dl*337 have not been tested for DNA degradation during their growth cycle.

*hr*6, *hr*cs13 (KARGER, HO, CASTIGLIA, FLINT, and WILLIAMS, unpublished work), *dl*338 (PILDER, LOGAN and SHENK, unpublished work) and other E1B 55K-specific deletion mutants (BABISS and GINSBERG, unpublished work) all grow poorly under nonpermissive conditions and appear to have similar phenotypes. They synthesize DNA normally, produce elevated levels of E2A mRNA and protein, and accumulate reduced levels of most mRNAs and proteins derived from the major late transcription unit. These mutants also fail to completely shut off the host cell. These observations indicate that the 55K polypeptide plays an important role during the lytic cycle of the virus, but as yet it is not possible to distinguish primary from secondary effects in the complex mutant phenotypes. They also demonstrate that viral DNA synthesis per se is not sufficient for normal expression of late gene functions.

In contrast to the other 55K-specific mutations, hr7 exhibits a DNA replication defect (STILLMAN 1983). DNA replication occurs in hr7-infected HeLa cells but its onset is delayed by about 10 h. Assuming that this defect is due to the hr7 mutation (and not a secondary alteration), E1B products also play a role in DNA replication. This is consistent with the observation that dl313 is unable to synthesize DNA in HeLa cells (JONES and SHENK 1979a). In this case, however, it was not possible to unambiguously assign the function of E1B, since E1A polypeptides were also altered. It is not yet clear why hr7 should display a defect in DNA replication while most other E1B mutants do not. Possibly, the 55K polypeptide contains an N-terminal replication function domain which is left intact in all the mutants except hr7 and dl313.

4.3 Region E1A and E1B Transforming Functions

Infection of rodent cells in vitro by group C adenoviruses follows a non- or semiproductive course, leading to the heritable transformation of a small percentage of the population. Morphologically, the transformants undergo a change from a fibroblastic to an epithelioid, rounded appearance, and they tend to grow as dense, multilayered foci or colonies. Little is known about the virus–cell interactions involved in the initiation and maintenance of adenoviral transformation, and identification of the viral gene products required is an obvious prerequisite to any functional analysis of the process.

Three main lines of evidence suggest that the information necessary and sufficient for this transformation is contained in the part of the genome comprising E1A and E1B. First, cell lines transformed by these viruses always contain and express the E1A and E1B regions and some possess only that sequence (GALLIMORE et al. 1974; SAMBROOK et al. 1974; FLINT et al. 1975, 1976). Second, transfection of rodent cells in vitro with DNA fragments has shown that left-end fragments alone are sufficient to bring about transformation (GRAHAM et al. 1974). A left-end fragment of 8% induces apparently complete transformation of primary rat kidney cells, while a smaller 4.4% fragment from the extreme left induces an only partially transformed phenotype (VAN DER EB 1977, 1979; HOUWELING et al. 1980). Third, it has been reported that hr mutants with mutations mapping to either E1A or E1B (see Figs. 1 and 2) are transformation defective (HARRISON et al. 1977; GRAHAM et al. 1978; JONES and SHENK 1979a; CARLOCK and JONES 1981; SOLNICK and ANDERSON 1982; Ho et al. 1982; BABISS et al. 1983a; CHINNADURAI 1983; BABISS, FISHER and GINSBERG, to be published).

Transformation experiments have been performed using *hr* mutants to define the roles of the various E1A and E1B gene products in the development of the transformed cell phenotype. GRAHAM et al. (1978) found that E1A mutants (e.g., *hr*1) did not transform rat embryo cells, but did induce abortive or incomplete transformation of baby rat kidney cells. E1B mutants (e.g., *hr*7) were unable to transform either of these cell types. One interpretation of this result is that E1A mutants are able to inititate transformation in kidney cells but the event is not fixed because the product of the mutated E1A gene is needed for maintenance of the transformed state. The E1B mutants are totally negative for induction in both cell types, and these mutations lie in a gene primarily required for initiation. However, an alternative explanation must be considered in view of the finding that an E1A product regulates expression of the other early regions (BERK et al. ,1979; JONES and SHENK 1979b). In this case the E1A gene product is required for initiation and would be required for normal expression of the E1B maintenance gene in transformed cells. Were this second hypothesis correct, E1B mutants might be expected to induce incomplete or abortive transformation of rat kidney cells, and this is not the case. A partial, but not abortive, transformation of 3Y1 rat cells by E1B *dl*313 has been reported (SHIR-OKI et al. 1981), although this has not been observed in embryonic cells by others (SHENK et al. 1979; WILLIAMS and HO, unpublished results). This apparent contradiction may be explained by the fact that 3Y1 is a permanent cell line, which is already immortalized. E1A mutant *dl*312 apparently does not transform these cells, but as yet no other *hr* mutants have been tested.

Several additional pieces of experimental evidence support the first of the two interpretations given above. First, E1A mutant *in*500, in which the insertion introduces a stop codon in the 13S message (nucleotides 1086–1088), is defective for transformation of baby rat kidney cells (CARLOCK and JONES 1981). This alteration apparently has no effect upon expression of E1B in nonpermissive HeLa cells. Second, E1A mutant *hr*440, which has an amber codon at positions 979–981 in the reading frame for the 13S mRNA, is also defective for transformation of baby rat kidney cells. Again, this mutation has no effect on control of expression of E1B products in HeLa or rat kidney cells. RUBEN et al. (1982) have reported that some lines of rat kidney cells, incompletely transformed by *hr*1, can eventually be established. These transformants remain fibroblastic, fail to grow in agar suspension, and apparently contain normal amounts of 21K and 55K E1B proteins. These observations all support the view that the E1B proteins, while they may be required, are not sufficient to maintain a fully transformed phenotype, and that functional E1A gene product is required for this. Interactions between these various products may be required for expression of a complete transformed phenotype.

The fact that E1B mutants *hr*6 and 7 do not transform any primary cell types and are defective for E1B 55K synthesis (LASSAM et al. 1979; ROSS et al. 1980b) strongly suggests that this protein is important for transformation. However, it has been shown that the left-end 8% fragment of Ad5 DNA is sufficient to transform rodent cells (GRAHAM et al. 1974) and, as expected, these fully transformed cells do not contain immunoprecipitable 55K protein (SCHRIER et al. 1979). This finding suggests that at least the C-terminal part of the 55K moiety is not needed for transformation, and is in apparent disagreement with the results obtained with the E1B *hr* mutants. ROWE and GRAHAM (1983) have provided an explanation for this seeming contradiction. They find that DNA from E1B mutants (*hr*6 and *hr*50) is capable of transforming both rat and hamster kidney cells. All the mutant-transformed lines examined lacked detectable 55K antigen and some also lacked the E1B 19K antigen. ROWE and GRAHAM (1983) suggest that the 55K protein is required for initiation of transformation when virus is used, but not when DNA is used. Their results also cast some doubt on the role of the E1B 19K protein in transformation. However, MATSUO et al. (1982) detected this antigen in all group C transformants examined and it is found in all rat kidney cell lines established by transformation with the left-end 8% DNA fragment (SCHRIER et al. 1979). Further, two pieces of genetic evidence support the view that the 19K product is required for transformation. First, insertional mutations introduced into the 19K region of cloned E1 DNA by the bacterial transposon *Tn*5 eliminate the transforming capacity of the DNA (MCKINNON et al. 1982). Second, viable, large-plaque mutants with mutations in the 19K sequence show reduced transforming capacity, whether virus or viral DNA is used (CHINNADURAI 1983; PILDER, LOGAN and SHENK, unpublished work). Although the role of the 19K protein in transformation is not yet clear, it does seem to be a good candidate for the adenovirus tumor-specific transplantation antigen (TSTA). It has been purified and is found as an integral part of the plasma membrane in infected and transformed cells (PERSSON et al. 1982). In this respect, it is of interest that *sub*315 (JONES and SHENK 1979a), which removes all of the 19K sequence and the N-terminal part of the 55K region, is totally unable to induce TSTA in rats (GALLIMORE and WILLIAMS 1982). The E1B mutants *hr*6 and *hr*7, which are defective for 55K protein production, show reduced but significant levels of TSTA induction compared with wild-type. This could mean either the 55K moiety also acts as a TSTA or that 55K activity is required for full 19K levels of activity.

ROWE and GRAHAM (1983) also examined the tumorigenicity of hamster cells transformed by DNA from E1B host-range mutants and discovered that all lines were able to form tumors, albeit in most cases with reduced efficiency and longer latent period. This suggests that the 55K protein, and perhaps also the 19K species, are not absolutely required for oncogenicity, although tests by others using rat cells and nude mice (JOCHEM et al. 1982; BERNARDS et al. 1982) suggest that oncogenicity of transformed cells is conferred or at least influenced by E1B products. Possible influences of cell type, spontaneous acquisition of tumorigenicity, and the test systems used, require that a fair degree of caution be imposed when interpreting or comparing these results.

The general conclusions to be drawn from all the above results are: (a) The protein translated from the E1A 13S mRNA is required directly for transformation and does not act indirectly by switching on E1B expression; and (b) the E1B 19K and 58K proteins are also required for transformation and the latter may be primarily involved in an initiation step. Does a viral gene product play a role in maintaining the phenotype of the transformed cell? This issue has been addressed with reference to mutants with a conditionally defective host-range and transforming capacity (HO et al. 1982). Mutants of Ad5 that are cold sensitive (*cs*) for growth in HeLa cells have been assigned to two complementation groups, one mapping physically to E1A and the other to E1B. All these *cs* mutants are defective for transformation of rat embryo cells. The E1A mutants are conditionally defective and transform with normal or elevated frequencies at 38.5 °C and with reduced frequencies at 35.5 °C and 32.5 °C. The E1B mutants transform poorly at all three temperatures. Of particular interest is the fact that *hr*1, which is not *cs* for infection of HeLa cells, is *cs* for transformation of rat embryo cells (Ho et al. 1982) and cells of the cloned rat line CREF (BABISS et al. 1983a). The results of temperature-shift experiments with *hr*1 and other E1A *hr cs* mutants support the view that the virus encodes a transformation maintenance function. Similarly, shift experiments with various *hr*1 and *hr cs*-transformed cell lines provide evidence that at least some phenotypic traits of transformants, such as anchorage independence, are under viral gene control. It has been demonstrated that this control is mediated directly by E1A and not via expression of E1B gene functions by showing a plasmid containing only the E1A genes of *hr*1 transforms with a *cs* phenotype (BABISS, FISHER and GINSBERG, to be published).

The *hr*1 mutation results in translation of a truncated protein from the E1A 13S mRNA. This protein, minus its C-terminal end, is still capable of some transforming capacity in that the mutant abortively transforms rat kidney cells and transforms rat embryo and CREF cells at high temperatures. This suggests that the transforming domain lies in its N-terminal region. The cold sensitivity of transformation might result either from the actual *hr*1 mutation, or from the string of new amino acids introduced as a result of frame-shift, or as the result of an additional mutation located in the 5′ half of the E1A sequence, outside the region sequenced in *hr*1 (RICCIARDI et al. 1981). A second-site mutation in this region would alter both the 13S and the 12S mRNAs. Removal of this smaller mRNA by site-specific mutagenesis (mutant *pm* 975) does not alter viral growth or early gene expression (MONTELL et al. 1982), but the 12S product might play a role in transformation since the mutant does show some reduction in transformation frequency (A. BERK, personal communication). These problems are now being addressed with the aid of mutants bearing deletions or insertions in the E1A coding region (BABISS, FISHER and GINSBERG, to be published). One of these constructed mutants, *dl*101, has a 5-bp deletion from nucleotides 1008 to 1012 (Fig. 1), which introduces a frame shift generating a stop codon at nucleotide 1241. Mutant *in*106 has a 16-bp insert starting at 1009, with a stop codon at position 1013 within the insert. A third, *dl*105, has a 69-bp deletion from 1002 to 1072. The last is defective for transformation at all temperatures, but the first two show a *cs* transforming phenotype on CREF similar to that of *hr*1. These results suggest that the *cs* phenotype of *hr*1 does not result from an additional mutation elsewhere on the genome. Further, since the string of missense amino acids found at the C-terminal end of the truncated proteins made by *hr*1 and *dl*101 is missing from *in*106 and all three mutants have the same transformation phenotype, it is unlikely that these added regions play a role in making the truncated E1A protein conditionally functional for transformation.

Most likely the cs phenotype of these mutants reflects a change in the ability of the E1A protein to interact or bind with another viral or cellular macromolecule. A precedent for such interaction in adenoviruses is provided by the interaction of the E1B 55K protein with a 53K cellular protein found in a variety of transformed cells (SARNOW et al. 1982b).

Clearly, the E1A protein is necessary for transformation, and the E1B proteins also contribute to the process. Transformation is a multistep and complex event and these various proteins may be responsible for determining different parts of the complete transformed phenotype, as is apparently the case for polyoma virus large and middle T antigens (RASSOULZADEGAN et al. 1982). In this respect, it is of great interest that both polyoma large T and adenovirus E1A gene functions effect immortalization of primary rat cells (RASSOULZADEGAN et al. 1982; HOUWELING et al. 1980) and complement the ability of human ras oncogenes to transform these cells (LAND et al. 1983; RULEY 1983).

4.4 Region E2B

Early region 2B lies between coordinates 11 and 75 on the adenovirus genome (Fig. 3). Previous mapping studies located leftward-reading transcripts corresponding to E4 and E2A but did not reveal any to the left of around position 62 on the genome (PHILIPSON et al. 1974; SHARP et al. 1974). The first hint of additional early transcription from the l-strand arose from physical mapping studies of ts group N mutants of type 5 adenovirus (GALOS et al. 1979; WILLIAMS et al. 1979). These mutations mapped between 18.5 and 22.0 units on the genome and transcription was found to occur at a low level early in infection from the l-strand sequence in this region and also further to the left between 11 and 14.5 map units. More detailed examination has revealed a collection of mRNAs whose main bodies are complementary to the l-strand sequence in this region (STILLMAN et al. 1981). The 5′ ends of these transcripts map around position 75, which was previously identified as the promoter region for the E2 transcription unit (KITCHINGMAN et al. 1977; BERK and SHARP 1978; CHOW et al. 1979). STILLMAN et al. (1981) renamed the E2 region E2A and called the regions with the newly identified transcripts E2B. These E2B transcripts have leader sequences from coordinates 76, 68, and 39, which are spliced to main bodies at 30, 26, and 23; these in turn extend to a common termination site at 11 units. In vitro translation of these mRNAs from E2B gives products with apparent molecular weights of 75K, 87K, and 105K daltons (STILLMAN et al. 1981). The 87K protein corresponds to the precursor to the 55K terminal protein (pTP) (STILLMAN et al. 1981), while the 105K protein is almost certainly the 140K polypeptide identified in vitro as a DNA polymerase (ENOMOTO et al. 1981; LICHY et al. 1982; STILLMAN et al. 1982; OSTROVE et al. 1983; STILLMAN and TAMANOI 1983).

Sequence determination across this region has recently been completed (GINGERAS et al. 1982; ALESTROM et al. 1982), and two large open reading frames have been mapped in it. One starts at nucleotide 10531 and ends at 8572 (28.8–23.5 units) and encodes the pTP (SMART and STILLMAN 1982). The other

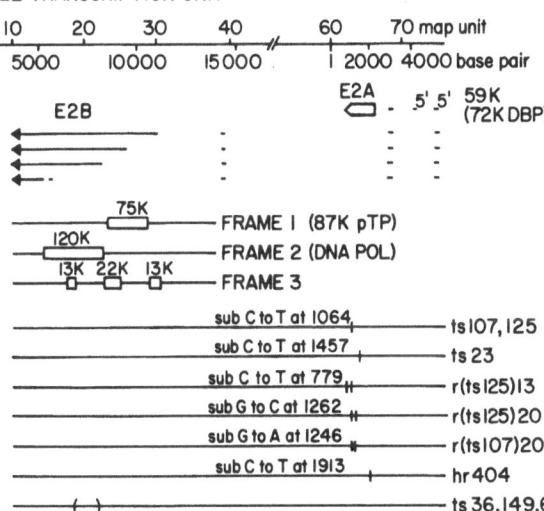

Fig. 3. Physical map of E2A and E2B mRNAs, coding regions, and mutations which alter E2 expression. The top of the figure positions the map in terms of map units and nucleotide sequence position. Sequence numbers for E2B are relative to the left end of the viral chromosome (ALESTROM et al. 1982; GINGERAS et al. 1982), and those for E2A are as designated by KRUIJER et al. 1981). Symbols and designations are as in Figs. 1 and 2 and the mutations are referenced in Table 1

large frame has an initial AUG at nucleotide 8354 and terminates at 5189 (22.9–14.2 units), which is sufficient to code for a protein of 120K. This almost certainly corresponds to the 105K polypeptide made by in vitro translation and to the 140K DNA polymerase mentioned above. Five other unidentified reading frames originate from the l-strand sequence in this region, with maximum polypeptide encoding capacities ranging from 10K to 31K. Thus, none is large enough to encode the 75K protein made in vitro (STILLMAN et al. 1981), and the origin and identity of this protein remain obscure. The r-strand in this region encodes the three major late leaders and the i-leader sequence, and there are four open reading frames capable of encoding proteins of 11.6K, 16.6K, 17.7K, and 14.4K. The region between coordinates 11 and 30, which appeared to be sparsely populated just a few years ago, is now a very crowded region and provides a good example of the elegant economy of the adenoviral genome.

The mutants of complementation group N map squarely in this region between coordinates 18.5 and 22.0 (GALOS et al. 1979). All four mutants of this group, ts36, ts37, ts69, and ts149, are negative for viral DNA synthesis at restrictive temperature (WILKIE et al. 1973; GINSBERG et al. 1974). The results of temperature-shift experiments suggest that the defect is primarily at the stage of initiation of synthesis but an elongation defect cannot be ruled out (GINSBERG et al. 1974; LEVINE et al. 1974). As might be expected, these mutants make no late mRNAs (BERGET et al. 1976) or late proteins (RUSSELL et al. 1974) at nonpermissive temperature, but the overall pattern of early RNA expression seems to be normal (BERGET et al. 1976). Expression of E1B mRNAs in cells at restrictive temperature has not yet been examined with these mutants. The map position of the mutations indicates that they lie within the reading frame encoding the 120K protein, and the fact that they induce a defect in DNA replication lends genetic support to the view that the 140K DNA polymerase is the product of that reading frame. In fact, based on the results of experiments designed to examine dominance interactions between mutants and wild-type viruses in mixed infections, it was predicted that the N gene mutations lie in the structural gene of a catalytic product (GALOS et al. 1979; WILLIAMS et al. 1979). A further prediction from these various results is, of course, that the 140K DNA polymerase induced in cells infected by ts36 will be thermosensitive. To date, no mutations have been mapped in the 80K pTP frame.

Elucidation of the components and steps involved in adenoviral DNA synthesis is now forging ahead as the result of the development of a soluble, in vitro replication system which uses exogenous adenoviral DNA template (CHALLBERG and KELLY 1979). The covalent linkage of pTP with dCMP, the 5′-terminal nucleotide in the new DNA strand, which constitutes the initial step in DNA replication, occurs in vitro (CHALLBERG et al. 1980, 1982; LICHY et al. 1981; TAMANOI and STILLMAN 1982). This complex provides a primer for nascent strand elongation by DNA polymerase. Formation of the pTP-dCMP product in vitro allows assay and subsequent purification of pTP from infected cell extracts (IKEDA et al. 1981). An alternative method for purification or replication proteins involves an in vitro complementation assay using DNA-defective ts mutants. For example, extracts of cells infected at restrictive temperature with ts36 or ts149 are defective for DNA replication in vitro, and these can be activated by addition of components purified from extracts of cells infected by wild-type virus (STILLMAN et al. 1982; OSTROVE et al. 1983). Mutants are complemented by a fraction which contains a complex of the 80K pTP and a 140K protein. The two can be separated and complementing activity is associated with the 140K protein, which possesses DNA polymerase activity. Similar co-purification of pTP with polymerase occurred when pTP-cDMP complex formation was used as the assay (IKEDA et al. 1981; ENOMOTO et al. 1981), but again separation of the two components has been achieved (LICHY et al. 1982). Both proteins are needed for initiation complex formation and for DNA synthesis and the polymerase probably catalyzes both attachment of dCMP to pTP and addition of nucleotides to the 3′-OH of the primer during elongation of the new DNA chain. Further genetic analysis will undoubtedly play an important role in determining just how these proteins function and interact to bring about DNA replication.

4.5   Region E2A

The E2A transcription unit lies between map coordinates 61.6 and 75, and the mRNA of the viral single-stranded DNA-binding protein (DBP), the only known translational product of the unit, is derived from the l-strand in this region of the genome. Sequence analysis of the DNA and mRNA (KRUIJER et al. 1981) showed that the DBP mRNA made early in infection has a body mapping between 66.5 and 61.6 map units with leaders at 75 and 68 units. In contrast, at late stages, leaders from 72 and 68 map units are attached to the main body although the same start codon and reading frame are used in both cases. The apparent molecular weight of the Ad5 and Ad2 DBPs is commonly determined at 72 000 daltons, while that calculated from the nucleotide sequence is 59 048 daltons. The discrepancy between these values may result from the coiled, nonglobular structure of the molecule imposed by its high proline content.

Functionally, the DBP is the best characterized of the adenoviral early proteins and much of what we know of it has come from studies of *ts* and other mutants whose mutations map in the *DBP* gene (Fig. 3). The protein binds efficiently, but not covalently, to single-stranded DNA (VAN DER VLIET and LEVINE 1973; VAN DER VLIET et al. 1978) and, to a lesser extent, to double-stranded DNA and DNA termini (FOWLKES et al. 1979). The DBPs of E2A prototype mutant *ts*125 (ENSINGER and GINSBERG 1972) and H2ND1*ts*23 (WILLIAMS and FRASER, unpublished results), are thermolabile (VAN DER VLIET et al. 1975). These mutants were found to be defective for viral DNA synthesis in cells infected at restrictive temperature, and it was proposed that the DBP is required for both initiation of synthesis and subsequent elogation (VAN DER VLIET and SUSSENBACH 1975; VAN DER VLIET et al. 1977; HORWITZ 1978). The use of in vitro systems for synthesis of adenovirus DNA provides evidence that DBP is not required for the initiation reaction, defined as formation of 80K–dCMP complex (see above; LICHY et al. 1981; CHALLBERG et al. 1982). Production of this complex takes place in the absence of purified wild-type DBP (ENOMOTO et al. 1981; IKEDA et al. 1981) and extracts from cells infected with *ts*125 make complex at restrictive temperature (CHALLBERG et al. 1982; FRIEFELD et al. 1983). Initial steps of the elongation reaction are reduced only slightly in mutant extracts incubated at restrictive temperature (FRIEFELD et al. 1983), suggesting either that DBP is not absolutely required or that a cellular DBP is sufficient for this process. As the elongation reaction progresses in vitro, however, there is an absolute requirement for viral DBP (FRIEFELD et al. 1983).

In addition to defective viral DNA replication, cells infected at restrictive temperature with *ts*125 contain elevated levels of early DBP mRNA and other early viral mRNAs, suggesting that the DBP autoregulates production of its own message (CARTER and BLANTON 1978b). Further studies have shown that overproduction probably results from increased stability of these mRNAs (BABICH and NEVINS 1981). The interaction could, of course, be indirect and mediated by another viral or cellular protein. Studies using *ts*125 have also suggested that the DBP represses transcription from E4 but not synthesis of its own message (NEVINS and WINKLER 1980). The mechanism of this repression is not yet understood but it seems to act at the level of initiation of transcription and does not involve premature termination of transcription. Again, it remains to be determined whether the DBP affects repression of E4 transcription directly or whether its effect is brought about by another viral or cellular gene product.

The DBP can be cleaved by chymotrypsin to give two fragments (KLEIN et al. 1979; LINNE and PHILIPSON 1980). The larger, 45K fragment, comprising the C-terminal region of the protein, carries the capacity for binding to single-stranded DNA, and is sufficient for complementation of *ts*125 DNA replication at restrictive temperature in vitro (ARIGA et al. 1980). A smaller, 34K subset of this C-terminal fragment is equally as active as the 72K molecule when used to complement mutant DNA synthesis in vitro (FRIEFELD et al. 1983). The 25K, phosphorylated, N-terminal fragment, which does not bind to single-stranded DNA, is not necessary for DNA synthesis in vitro. Thus, both biochemical and genetic approaches assign the functional domain of the DBP required for DNA synthesis to the C-terminal region of the molecule.

In addition to the *ts* mutants discussed above, another class of mutants which alter the host range of Ad2 and Ad5 on monkey cells has been isolated and mapped to this gene (KLESSIG and GRODZICKER 1979; KRUIJER et al. 1981). Normally, human adenoviruses grow poorly in monkey cells, although early functions are expressed and viral DNA is replicated efficiently. The levels of some late viral mRNAs and proteins are reduced in these cells (ERON et al. 1975; KLESSIG and ANDERSON 1975). This is especially true of fiber gene expression. Its mRNA levels are down some 5- to 20-fold, about half of the fiber mRNA that is made in these cells is defectively spliced (KLESSIG and CHOW 1980), and protein levels are reduced 100- to 1000-fold. The observed RNA defects alone may be sufficient to reduce fiber translation to the low levels seen, but further restriction may be imposed by monkey cell translational components. In contrast to wild-type virus, the DBP *hr* mutants display correct splicing of the fiber mRNA (KLESSIG and CHOW 1980), make normal amounts of all late proteins, and grow efficiently in monkey cells (KLESSIG and GRODZICKER 1979). On the basis of these findings, KLESSIG and GRODZICKER have postulated that DBP normally interacts with certain human cell factors involved in late viral gene expression (perhaps cell factors involved in RNA splicing) but that it cannot interact efficiently with the equivalent components in monkey cells. The *hr* mutations alter the DBP so that it can interact effectively with the monkey cell factors to allow correct RNA processing and late viral protein synthesis. Whatever the actual mechanism by which DBP acts, these mutations reside in the region of the gene encoding the N-terminal part of the DBP and it is not unreasonable to suggest that this region comprises a separate, late-functional domain of the protein.

The host-range block in monkey cells can also be overcome by co-infection with SV40, and it has been shown through use of Ad2$^+$ND$_1$ hybrids and deletion mutants of SV40 that the helper activity resides in the C-terminal region of the SV40 large T antigen (KELLY and LEWIS 1973; LEWIS et al. 1973; COLE et al. 1979). It is therefore possible that T antigen and DBP function in a similar fashion to permit efficient expression of late adenoviral genes in monkey cells. Further evidence for possible functional relatedness of T and DBP arises from the observations that SV40 complements (or perhaps suppresses) the *ts*125 defect in monkey cells co-infected at restrictive temperature and allows adenoviral DNA synthesis to proceed (WILLIAMS et al. 1974; LEVINE et al. 1974; RABEK et al. 1981; GOLDMAN et al. 1981). It is of interest that SV40 efficiently complements H5*ts*19 (WILLIAMS et al. 1974), a late-function, DNA-positive mutant, whose mutation maps by marker rescue between 63.6 and 68.0 map units (Fig. 6). There is therefore a very good chance that this mutation also resides in the DBP structural gene and alters the N-terminal, late-function domain of the protein.

The overall picture emerging from these observations is that the adenoviral DBP is a versatile, multifunctional protein with a number of separate functional domains. While additional DBP mutants will undoubtedly improve our knowledge of functional organization in this protein, a parallel genetic approach promises to be very valuable for the analysis of structure and function in DBP. This involves isolation and characterization of temperature-independent revertants of *ts* mutants which bear second-site suppressor mutations (NICOLAS et al. 1981, 1982; LOGAN et al. 1981; KRUIJER et al. 1983). Revertants of four genotypic groups comprising two phenotypic classes have been isolated (Fig. 3). Group I revertants are true revertants in that they possess a wild-type nucleotide sequence across the DBP region. Group II revertants, typified by r13, possess a second-site mutation in the DBP gene at nucleotide 779, resulting in substitution of the histidine at position 508 to a tyrosine. In human cells, these mutants behave phenotypically as wild-type, except that they overproduce a stable DBP in HeLa cells but not in 293 or CV1 monkey cells (NICOLAS et al. 1982). The host-cell specificity might mean that the DBP interacts with a cellular factor to bring about autoregulation of its own synthesis. Group III and group IV revertants possess second-site mutations at nucleotides 1246 and 1262 respectively and comprise the second phenotypic class. These revertants display host-range temperature-sensitivity in that they are *ts* for growth in 293 cells and behave as wild-type in HeLa cells (NICOLAS et al. 1981). This property is again consistent with the hypothesis that DBP interacts with cellular factors to elicit certain facets of its multifunctional character.

4.6 Region E2A and E2B Transforming Functions

As discussed above, the adenoviral genetic information required and sufficient to induce transformation of rodent cells lies in the E1A/E1B region of the genome. Thus, one would predict that gene products encoded elsewhere in the genome do not play a direct role in this process. Nevertheless, it has been found that mutations in both the *N* (polymerase) gene and the *DBP* gene do influence transformation by Ad5.

All four mutants of the *N* gene, *ts*36, *ts*37, *ts*69, and *ts*149, have been found to transform rat embryo cells at permissive temperature with wild-type frequency, but to transform with reduced frequency at restrictive temperature (WILLIAMS et al. 1974, 1979). GINSBERG et al. (1974) reported that *ts*149 is not thermosensitive for transformation, but WILLIAMS et al. (1979) have been unable to repeat that result. The results of temperature-shift experiments show that an initiation step in transformation is involved; cells held at permissive temperature for 2 days prior to shift-up to restrictive temperature are transformed at wild-type frequencies. Three pieces of evidence support the view that the transformation phenotype of these mutants results from the *N* gene mutations and not from independent *ts* mutations located in the left-end transforming region. First, all four *N* mutants, which were independently isolated, display the phenotype. Second, a spontaneous revertant of *ts*36, *ts*$^+$ for lytic growth, transforms with wild-type frequency (WILLIAMS, unpublished results). Third, left-end DNA fragments from $ts36$, which exclude the N gene, transform with wild-type frequency (J. ARRAND, personal communication).

A variety of hypotheses can be proposed to explain the role of the N gene product in the events leading to transformation. The DNA polymerase mutations could, for example, prevent transformation by preventing viral DNA synthesis. However, viral DNA synthesis per se does not seem to be required for transformation, since DNA-negative DBP mutants (see below) transform effectively at restrictive temperature. Alternatively, the N gene product may play a role either in the processing of viral DNA prior to integration into the cell genome or in one or more of the events of integration. Presumably such a function is not required when DNA fragments are used in transformation, since left-end fragments excluding the N gene are capable of transforming cells. Lastly, it is possible that the N gene product plays a role in regulating the levels or activities of E1A or E1B transforming gene products during infection of rodent cells. Whatever the explanation, the effect of the N gene is likely to be an indirect one, manifest only when virus is used for transformation.

In contrast to the temperature-dependent effect of E2B mutations, ts E2A mutants display elevated transformation frequencies at both permissive and restrictive temperatures, but more markedly at the latter (GINSBERG et al. 1974; WILLIAMS et al. 1974). Revertants of these mutants with second-site mutations in the DBP gene also transform at high frequency, but true ts^+ revertants transform with wild-type frequency (LOGAN et al. 1981; CARTER et al. 1982). This latter result, plus the fact that left-end DNA fragments from $ts107$ transform with wild-type frequency (LOGAN et al. 1981), suggests that the transformation phenotype does not result from a second-site mutation in the transforming region. As with the N gene product, the role of the multifunctional DBP in transformation is undoubtedly indirect and explanation of the effect is open to conjecture. Possibly the DBP influences integration of viral DNA into cellular DNA. In this respect it may be significant that many cells transformed by $ts125$ at restrictive temperature carry sequence representative of most of the viral genome, as opposed to only left-end sequence as in many wild-type transformants (MAYER and GINSBERG 1977). Perhaps of more significance for enhanced transformation is the fact that DBP reduces stability of E1A and E1B messages in human cells (BABICH and NEVINS 1981; see above). At restrictive temperatures, less DBP is produced by ts mutants and levels of E1 mRNAs rise. Were this to occur also during infection of rodent cells at restrictive temperature, it would provide a logical explanation for the high transformation frequency. Finally, it must also be considered that the DBP might regulate the level of the E2B gene product which, in turn, affects transformation.

4.7 Region E3

The E3 transcription unit lies between 76 and 86 map units (Fig. 4). It encodes a variety of different mRNA species, which differ both in their splicing patterns and in poly A site utilization (BERK and SHARP 1978; CHOW et al. 1979a).

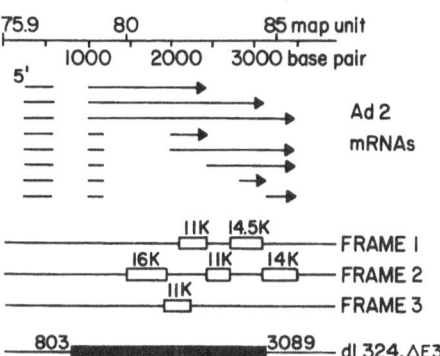

E3 TRANSCRIPTION UNIT

Fig. 4. Physical map of E3 mRNAs, coding regions, and mutations which alter E3 expression. The *top* of the figure positions the map in terms of map units and nucleotide sequence position. Sequence numbers are as designated by HERISSE et al. (1980) and HERISSE and GALIBERT (1981). Symbols and designations are as in Fig. 1, and the mutations are referenced in Table 1

Open coding regions exist in all three reading frames (HERISSE et al. 1980; HERISSE and GALIBERT 1981). E3 encodes a glycopolypeptide with an apparent molecular weight of 19K daltons (PERSSON et al. 1979; ROSS et al. 1980a). SIGNAS et al. (1982) reported that the glycoprotein binds heavy chains of class I transplantation antigens from human and mouse cells.

A wide variety of E3 deletion and substitution mutants has been described, and they are all viable. The two largest deletions in this area (which are identical) are shown in diagram form in Fig. 4. *dl*324 (THIMMAPPAYA et al. 1982) and ΔE3 (BERKNER and SHARP, to be published) both lack 2286 bp, deleting all of the E3-specific open coding regions. Both viruses grow well, indicating that E3 gene products are not required for growth of adenoviruses in cultured cells. In fact, BERKNER and SHARP (to be published) found that in mixed infections E3 outgrows its parent or wild-type virus, which has an intact E3 region.

The fact that E3 products are nonessential for viral growth in cultured cell lines does not mean that this gene fails to perform a useful function for the virus in nature. Identification of this function may have to await the development of a whole-animal model for adenovirus infection.

4.8 Region E4

The E4 transcription unit lies between 91 and 100 map units (Fig. 5). As with most other adenovirus transcription units, it encodes a variety of different mRNAs (BERK and SHARP 1978; CHOW et al. 1979) and contains open coding regions in all three reading frames (HERISSE et al. 1981; GINGERAS et al. 1982). There is also a coding region beyond the E4 poly A site near 91 map units, and several others exist on the opposite strand. A variety of polypeptides has been assigned to E4-specific mRNAs by in vitro translation of selected RNAs (e.g., TIGGES and RASKAS 1982), and three polypeptides have been identified in vivo and unambiguously assigned to E4 coding regions. These polypeptides have apparent molecular weights of 25K (SARNOW et al., to be published), 14K (DOWNEY et al. 1983), and 11K daltons (SARNOW et al. 1982a). The 25K poly-

E4 TRANSCRIPTION UNIT

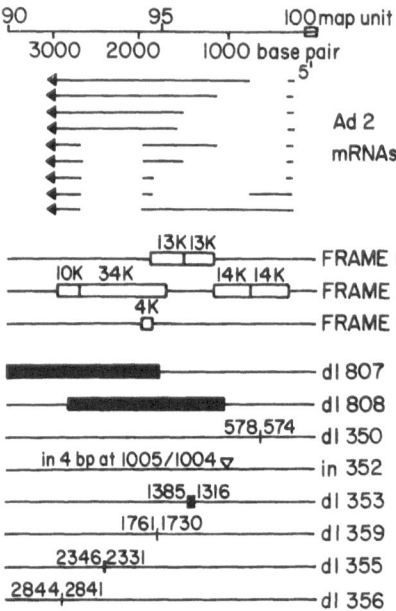

Fig. 5. Physical map of E4 mRNAs, coding regions, and mutations which alter E4 expression. The *top* of the figure positions the map in terms of map units and nucleotide sequence position relative to the right end of the viral chromosome (GINGERAS et al. 1982). Symbols and designations are as in Fig. 1, and the mutations are referenced in Table 1

peptide exists at least partially in a complex with the E1B 55K polypeptide, and the 11K species is associated with the nuclear matrix.

CHALLBERG and KETNER (1981) have isolated several naturally arising defective viruses which contain deletions within the E4 region (*dl*807 and 808, Fig. 5). The viruses were propagated with the aid of a *ts* helper virus. It was possible to separate *dl*807 from its helper by exploiting density differences (*dl*807 virions are less dense than helper virions). The growth characteristics of purified *dl*807 virions (lacking the region from 82.5 to 95 map units, which includes E3, fiber, and E4 genes) were then analyzed. Viral DNA was synthesized normally in *dl*807-infected cells while some, but not all, late proteins were produced in reduced amounts.

HALBERT, CUTT and SHENK (unpublished work) have produced a series of small deletion and insertion mutants which carry alterations within individual E4 coding regions (*dl*350–359, Fig. 5). All the mutations were initially constructed in plasmids and then rebuilt into intact viral chromosomes. All the variants are viable and only one, *dl*355, grows more poorly than the wild-type parent, growing to approximately 50-fold reduced titers. This deletion is located within the 34K coding region, which elaborates the E4 polypeptide with an apparent molecular weight of 25K. This is the polypeptide known to complex with the E1B 55K polypeptide (SARNOW et al., to be published). Interestingly, the *dl*355 phenotype appears very similar to that of E1B 55K mutants. DNA replication is normal, E2A mRNA and protein are elevated, late mRNAs and proteins are reduced, and host cell metabolism is not completely shut off.

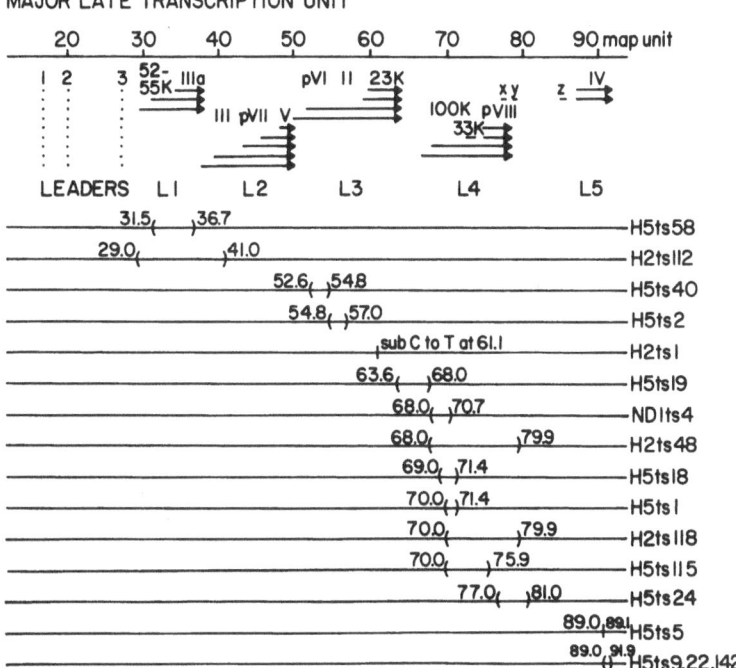

MAJOR LATE TRANSCRIPTION UNIT

Fig. 6. Physical map of the major late transcription unit mRNAs, coding regions, and mutations which alter their expression. The *top* of the figure positions the map in terms of map units. Symbols and designations are as in Figs. 1 and 2, except that coding regions are indicated by the name of the polypeptide product. References for the mutations are provided in Table 1

WEINBERG and KETNER (1983) have recently described the production of a monkey kidney cell line (Vero) which expresses region E4 function and can complement defective deletion mutants. This line should dramatically facilitate further analysis of region E4 function.

4.9 Major Late Transcription Unit

The major late transcription unit occupies a large part of the genome and, late in infection, after DNA replication, transcription starts at 16.3 on the r-strand (EVANS et al. 1977; ZIFF and EVANS 1978) and continues to the right end of the chromosome (FRASER et al. 1979). The large transcripts derived from this region generate five blocks of overlapping mRNAs and each of the mRNAs in a block has a common 3′ terminus (Fig. 6) (CHOW et al. 1977; NEVINS and DARNELL 1978; ZIFF and FRASER 1978; McGROGAN and RASKAS 1978; ZIFF and EVANS 1978; BERGET and SHARP 1979). In all, 18 late mRNAs have been mapped and each apparently bears the same tripartite leader sequence of 200 nucleotides generated from sequences around 16.3, 19.6, and 26.6 units on the

r-strand (BERGET et al. 1977; KLESSIG 1977; CHOW et al. 1977; DUNN and HAS-
SELL 1977; AKUSJARVI and PETTERSSON 1979; ZAIN et al. 1979). Transcription
from the late promoter at 16.3 also occurs early in infection, prior to viral
DNA replication, but it stops midway across the genome and does not proceed
to the end of the unit as in the late phase of infection (CHOW et al. 1979 b;
LEWIS and MATHEWS 1980; NEVINS and WILSON 1981). Further, there is a change
in the mode of splicing of L1 RNAs, so that different mRNAs are generated
from the same DNA sequence (THOMAS and MATHEWS 1980; NEVINS and WIL-
SON 1981).

To date, 11 late proteins have been identified and polypeptides have been
assigned to each of the five blocks of late mRNAs (Fig. 6). Three of these,
the 23K (L3), 33K, and 100K (both L4) are nonstructural proteins; the others
are all structural components of the virus particle (see RUSSELL and SKEKEL
1972; ANDERSON et al. 1973; EVERITT et al. 1973; ISHIBASHI and MAIZEL 1974;
LEWIS et al. 1977; AXELROD 1978; GAMBKE and DEPPERT 1983; OOSTEROM-
DRAGON and ANDERSON 1983; YEH-KAI et al. 1983). Very little is known about
the factors which control the production of the late viral proteins or how these
proteins interact with each other and with cellular components to guarantee
productive infection and cause cell death. So far, mutants have told us little
or nothing about these processes in adenovirus infections.

Many late-function *ts* mutants of types 2 and 5 with an array of phenotypes
have been isolated. Representatives have been mapped with certainty to four
of the late mRNA blocks by either intertypic recombinant analysis or marker
rescue, and in one case by nucleotide sequence analysis (Fig. 6). However, it
should be pointed out that although these various mutations map in the regions
of late viral genes, few have been assigned unambiguously to the reading frames
for particular proteins. Where these are close together, and, where frames overlap, a given mutation might alter more than one protein.

A large number of *ts* mutations have been mapped in the region of the
100K gene (66.5–73.2 units) and studies with some of these mutants (e.g., H5*ts*1
and H5*ts*115; Fig. 6) suggest an important role for this protein in the early
stages of virion assembly. These mutants show reduced levels of 100K or insta-
bility of 100K at restrictive temperature, defective assembly of hexon monomers
into trimers, and defective transport of hexon into the nucleus (RUSSELL et al.
1972, 1974; STINSKY and GINSBERG 1974; LEBOWITZ and HORWITZ 1975; KAUFF-
MAN and GINSBERG 1976; OOSTEROM-DRAGON and GINSBERG 1981; WILLIAMS
and SKLADANY, unpublished results). These results suggest that 100K and hexon
polypeptides interact in the cytoplasm of the infected cell to effect trimerization,
and agree with the results of serological analysis which show that 100K-specific
antisera co-precipitate 100K and hexon from infected cell extracts and that
the complexed hexon is the monomeric form of the protein (OOSTEROM-DRAGON
and GINSBERG 1980; GAMBKE and DEPPERT 1983).

Most of the mutants mapping to the 100K region fall into one large complementation group, but at least six additional complementation groups map
there (WILLIAMS et al. 1974; FROST and WILLIAMS 1978). All but one of these
map within the limits 58.0 and 71.4 units (R.S. GALOS and J. WILLIAMS,
unpublished results) and thus, map exclusively in the 100K reading frame. The

exception is *ts*17, which maps between 68.0 and 73.6 units, and since the *N*-terminal part of the 33K coding region overlaps the C-terminal part of the 100K frame by about 300 nucleotides (OOSTEROM-DRAGON and ANDERSON 1983) that mutation might also alter the 33K protein. The positive complementation with these various 100K mutants presumably results from intracistronic complementation and suggests either that the 100K acts in a multimeric form or that two or more 100K molecules cooperate to assemble and/or transport hexon trimers. One mutant in one of these groups, H5*ts*18 (69.0–71.4 units), displays a different phenotype from all the others and apparently has no effect upon hexon transport, although virus particles do not form (RUSSELL et al. 1972). In addition, this mutant, unlike all the other 100K mutants, is defective for induction of interferon in chick cells at restrictive temperature (USTACELEBI and WILLIAMS 1972). These results are consistent with the view that the 100K protein performs some other function(s) in the viral life cycle. It would be surprising if this large, very abundant, nonstructural protein did not.

Mutations in a number of other viral genes result in blocks at certain stages in virion assembly (RUSSELL et al. 1972; WEBER 1976; EDVARDSSON et al. 1978; D'HALLUIN et al. 1980; CHEE-SHEUING and GINSBERG 1982). Mutants mapped to the fiber gene, e.g., H5*ts*, 5, 9, 22, and 142 (Fig. 6) and H2*ts*125 (D'HALLUIN et al. 1980, 1982), all produce reduced levels of fiber protein, and make only empty or intermediate particles at nonpermissive temperature. The most complete and satisfying study to date has been carried out with H5*ts*142 (CHEE-SHEUNG and GINSBERG 1982). This mutant complements members of two other fiber-defective complementation groups, namely H5*ts*5 and H5*ts*22, but fails to complement a member of the third group, H5*ts*9 (see WILLIAMS and USTACE-LEBI 1971; RUSSELL et al. 1972; CHEE-SHEUING and GINSBERG 1982). All of these mutants map squarely in the fiber gene and complementation is presumably intracistronic, resulting from cooperative interaction of the different mutant polypeptides in the fiber trimers. H5*ts*142 makes nonglycosylated fibers at restrictive temperature and these are apparently assembled into trimers which are immunologically nonreactive, but which can be further constructed into virions on temperature shift-down. In addition, empty particles of low buoyant density are made at high temperature. On shift-down to permissive temperature, particles of intermediate density, and some with the density of complete virions, are formed. The latter contain complete complements of DNA and core proteins but, as with so-called young virions, they apparently contain nonprocessed proteins, pVI, pVII, and pVIII. The intermediate particles possess subgenomic amounts of encapsidated DNA, and hybridization studies show that they possess a significant preponderance of left-end sequence. The findings are consistent with the idea that the genome inserts left-end first into preformed, empty particles and agree with the findings of others (DANIELL 1976; TIBBETTS 1977; HAMMARSKJOLD and WINBERG 1980; HEARING and SHENK 1983a). Because there is a huge excess pool of major capsid proteins in the infected cell, and only a small fraction of this enters virions, it is difficult to determine the precursor–product relationship between the different varieties of particles with certainty. Nevertheless, the results agree with the general lineage of empty–intermediate–complete particle suggested by previous biophysical/biochemical studies.

Mutants which map in the region of the *IIIa* gene also accumulate assembly intermediates (EDVARDSSON et al. 1978; D'HALLUIN et al. 1978). H5*ts*58 accumulates intermediate particles at restrictive temperature but these are apparently thermolabile and not used for virion assembly upon temperature shift-down. H2*ts*112 similarly forms intermediate particles of light buoyant density at high temperature and these apparently contain a subgenomic fragment of DNA but may lack core proteins pVII and V. On temperature shift-down, some of these intermediate forms apparently do mature into complete virions, supporting the view that the intermediate density particles correspond to intermediate stages in the adenovirus assembly line.

A signal feature of the late phase of adenoviral assembly is the processing by proteolytic cleavage of certain virion polypeptides (ANDERSON et al. 1973; ISHIBASHI and MAIZEL 1974; EDVARDSSON et al. 1976). This event may be essential for the infectivity of the completed particle. Ad2*ts*1 is defective for cleavage of virion polypeptides pVI, pVII, and pVIII, and produces normal amounts of noninfectious particles at restrictive temperature (WEBER 1976; MIRZA and WEBER 1977). The mutant is also defective for cleavage of 80K pTP to the 55K form (CHALLBERG and KELLY 1981; STILLMAN et al. 1981). BHATTI and WEBER (1979) describe the appearance of a PVII-specific, neutral, chymotrypsin-like protease activity in Ad2-infected cells and young virions at later stages of infection. The activity is absent from cells infected with H2*ts*1 at restrictive temperature and is not present in the virions made by the mutant at this temperature. The *ts*1 mutation has been mapped by marker rescue and sequence analysis to a region of the genome between the hexon gene and the 72K *DBP* gene (YEH-KAI et al. 1983). Nucleotide sequence analysis of this region previously disclosed an r-strand-specific open reading frame with capacity to encode a 23K polypeptide between 60 and 61.7 units (KRUIJER et al. 1980; AKUSJARVI et al. 1981). All of this provides strong genetic evidence, but does not prove, that the 23K polypeptide is a viral protease. The enzyme activity has been described as chymotrypsin-like and the molecular weight of the *ts*1 protein is very similar to that of chymotrypsin. Nevertheless, the 23K protein could be required either directly or indirectly for either the induction of synthesis or the activation of a cellular protease. Ultimate proof that the protease is virus-coded will require purification of the enzyme and comparison of its sequence with that of the 23K protein gene and/or demonstration that the mutant enzyme is thermosensitive.

Despite the fact that an abundance of late *ts* mutants have been isolated, the full potential of mutants for analysis of adenovirus assembly has yet to be realized. Extension of the genetic approach will probably depend upon development of practicable in vitro complementation systems.

4.10 Virus-Associated RNAs

The virus-associated RNAs (VA RNAs) are small RNA polymerase III-transcribed species (REICH et al. 1966; WEINMANN et al. 1974). These RNAs are encoded by two genes (VAI and VAII), which are located at about 30 map

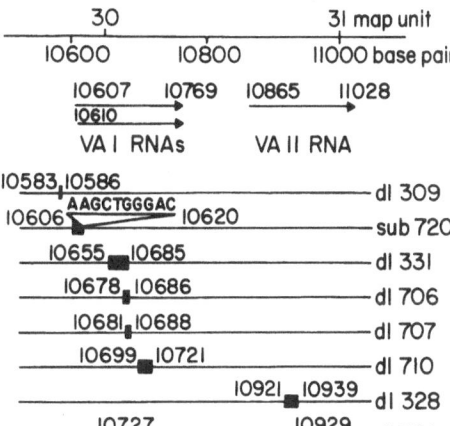

Fig. 7. Physical map of the region encoding VA RNAs and mutations which alter their expression. The *top* of the figure positions the map in terms of map units and nucleotide sequence position relative to the left end of the viral chromosome (GINGERAS et al. 1982)

units on the viral chromosome (Fig. 7) (MATHEWS 1975; PETTERSSON and PHILIP-SON 1975). There are two species of VAI RNA, which differ by three nucleotides at their 5' ends. The VAI and II RNAs are both about 160 nucleotides in length, and they are produced in remarkably large amounts late in infection.

Mutational analysis of cloned VA genes has established that, like other polymerase III-transcribed genes, they contain intragenic control regions (FOWLKES and SHENK 1980; GUILFOYLE and WEINMANN 1981). A viral mutant carrying a lesion in the VAI 5'-flanking region (*dl*309, Fig. 7) produces only one of the two VAI species, indicating that this region also influences initiation by RNA polymerase III (THIMMAPPAYA et al. 1979). The mutant grows as well as the wild-type virus, suggesting that multiple VAI species are not required for optimal growth.

*dl*331 (VAI$^-$/VAII$^+$) and *dl*328 (VAI$^+$/VAII$^-$) (Fig. 7) each fail to produce one of the VA RNA species (THIMMAPPAYA et al. 1982), since they carry deletions within the VA intragenic control regions. *dl*328 grows as well as the wild-type virus, suggesting that VAII RNA is not important for virus growth in cell culture. *dl*331 grows at a slower rate and reaches a 20-fold reduced yield compared with its parent. Although *dl*331 makes normal amounts of late mRNAs, they are translated at reduced efficiency (THIMMAPPAYA et al. 1982). Therefore, VAI RNA is required for efficient translation of viral mRNAs at late times after infection. Further analyses suggest that VAI RNA functions early during the initiation of translation (SCHNEIDER, WEINBERGER and SHENK, unpublished work). Possibly VAI RNA serves to properly localize viral mRNAs in the cytoplasmic compartment, perhaps by attaching them to the cytoskeleton where they can be translated (CERVERA et al. 1981) or to cellular structures which not only facilitate initiation but also allow proper cellular localization of newly synthesized polypeptides. Alternatively, the RNA could direct one or more initiation factors to viral mRNAs. Both models are attractive, since they provide a rationale for the large amount of VA RNAs present in infected cells. In one case, the RNA performs a structural role; in the other, it outcompetes, by virtue of numbers, a presumptive cellular counterpart.

BHAT and THIMMAPPAYA (personal communication) have produced a series of mutant viruses carrying alterations in the VAI RNA gene (*dl*706, 707, 710, 721, and *sub*720, Fig. 7). These viruses all lack the VAII species, so VA function must be provided by the altered VAI product. Since all the viruses, with the exception of *dl*721, grow quite well, it is reasonable to conclude the segments the modified VAI molecules lack are not essential to VAI RNA function. The deletion present in *dl*721 fuses the 5' end of the VAI to the 3' end of the VAII species. This mutant grows to a 20-fold reduced yield compared with its parent.

4.11 Control Regions

A variety of mutants has been generated to investigate transcriptional control regions, the leader sequences present on mRNAs encoded by the major late transcription unit, and the *cis*-acting viral packaging sequence.

The E1A transcriptional control region contains an element with enhancer properties located between nucleotide sequence positions 194 and 305. *in*340-2 (Fig. 1) lacks this sequence, and both its rate of transcription and the steady-state levels of E1A mRNAs are reduced about 20-fold (HEARING and SHENK 1983a). The enhancer element includes repeated core sequences. *in*340-A5 and B1 each lack one copy of this core and their transcription rates are reduced about 3-fold. A large number of deletion mutants have been constructed which lack sequences between the enhancer region and the start site for E1A transcription (OSBORNE et al. 1982; HEARING and SHENK 1983b). As has been shown for a variety of transcription units, deletion of the TATA consensus sequence reduced transcription about 5-fold, and deletions between the TATA box and start site displaced the initiation site downstream by about the size of the mutation.

Adenovirus constructions have been produced in which the major late control region has been inserted into the E1A transcription unit replacing portions of the normal E1A control region. SOLNICK (1983) constructed a variant (R103) carrying a 440 base-pair segment from the major late control region in this position. The insert functioned efficiently, increasing the levels of E1A mRNA 20-fold at late times after infection. Similar increases in E1A transcription have been observed by inserting a segment extending from -122 to $+33$ (relative to the major late start site at $+1$), while a segment including -52 to $+33$ was less active (*sub*364 and *sub*365, LOGAN and SHENK, unpublished work). These results suggest that a component required for optimal activity of the major late control region lies between -52 and -122 relative to the transcriptional start site.

At the other end of the transcription unit, the sequence AAUAAA has been shown to constitute at least part of the polyadenylation signal in many eukaryotic mRNAs (PROUDFOOT and BROWNLEE 1974; FITZGERALD and SHENK 1981). Since transcription proceeds beyond the polyadenylation signal, an endonucleolytic cleavage probably occurs to generate the substrate for poly A addition. MONTELL et al. (1983) have introduced a U-to-G transversion in the E1A AAUAAA sequence (*pm*1610, Fig. 1). This mutation decreases the efficiency with which the initial transcript is cleaved. However, RNA molecules which are cleaved are properly polyadenylated. These results suggest that the AAUAAA sequence signals cleavage but is not required for poly A addition.

Control sequences may also play a role in the efficiency with which mRNAs are translated in virus-infected cells. The R3 variant of SOLNICK (1981), which was discussed above, produced an mRNA that was translated efficiently in vitro but inefficiently in vivo compared with normal late mRNAs. THUMMEL et al. (1983) have found that hybrid adeno-SV40 mRNAs are translated at very different efficiencies, depending on the precise composition of the mRNAs. The tripartite leader, which is appended to all mRNAs produced from the major late transcription unit, is a logical candidate for a sequence that could modulate the efficiency with which an mRNA is translated within an infected cell. Experiments are in progress in a number of laboratories to evaluate the function of this sequence.

Adenoviruses contain *cis*-acting packaging sequences near the left end of the viral chromosome (TIBBETTS 1977; HAMMARSKJOLD and WINBERG 1980). An essential portion of the Ad16 sequence is located between 268 and 378 base pairs from the left end of the Ad16 genome (WINBERG and HAMMARSKJOLD, personal communication). A packaging sequence has been identified on the Ad5 chromosome within nucleotide sequence position 194–305. *in*340-2 (Fig. 1) lacks this sequence at its left end (HEARING and SHENK 1983a), but carries the segment inserted at the equivalent position at the right end of the chromosome. Interestingly, the packaging sequence lies on top of or is very close to the E1A enhancer element. The two regulatory elements may influence each other. Active utilization of the enhancer element could delay the onset of packaging; the onset of packaging could in turn decrease E1A transcription. More precise mapping of these two elements will better suggest how likely each is to influence the other's action.

5 Limitations of Genetic Analysis

The production and analysis of mutants has proven very effective in the identification of functions on the viral chromosome.

Very few technical problems limit our ability to introduce mutations at desired locations on the viral chromosome. The most serious problem in this area remains the propagation of defective mutants. However, it is likely that both additional cell lines which contain and express viral genes (BABISS et al. 1983b; WEINBERG and KETNER, 1983) and lines which carry suppressor tRNA genes (HUDZIAK et al. 1982) will be available in the near future. This should provide the means of propagating both nonconditionally defective and conditionally defective nonsense mutations, and overcome the last technical limitation to adenovirus genetic analysis.

One last problem exists, and is likely to remain. This is to distinguish primary from secondary effects of mutations. Not surprisingly, adenovirus gene expression is regulated by a complex interplay of gene functions. A mutation that modifies just one function can lead to a complex phenotype, many of whose components are indirect effects of the alteration under study. This problem, of course, is not unique to adenovirus. There is no easy solution, only thorough analysis and cautious interpretation.

Acknowledgments. We acknowledge the competent secretarial assistance of Ms. KATHLEEN DONNELLY and Ms. KATHRYN GALLIGAN. Unpublished work quoted from the authors' laboratories was supported by a grant from the American Cancer Society (MV-45) to THOMAS SHENK and a grant from the National Institutes of Health (CA-21375) to JIM WILLIAMS.

References

Aiello L, Guilfoyle R, Huebner K, Weinmann R (1979) Adenovirus 5 DNA sequences transcribed in transformed human embryo kidney cells (HEK-Ad-5 or 293). Virology 94:460–469

Akusjarvi G, Pettersson U (1979) Sequence analysis of adenovirus DNA: complete nucleotide sequence of the spliced 5' noncoding region of adenovirus 2 hexon messenger RNA. Cell 16:841–850

Akusjarvi G, Zabielski J, Perricaudet M, Pettersson U (1981) The sequence of the 3' non-coding region of the hexon mRNA discloses a novel adenovirus gene. Nucleic Acids Res 9:1–17

Alestrom P, Akusjarvi G, Pettersson M, Pettersson U (1982) DNA sequence analysis of the region encoding the terminal protein and the hypothetical N-gene product of adenovirus type 2. J Biol Chem 257:13492–13498

Anderson CW, Baum PR, Gesteland RF (1973) The processing of adenovirus 2 – induced proteins. J Virol 12:241–252

Ariga H, Klein H, Levine AJ, Horwitz MS (1980) A cleavage product of the adenoviral DNA-binding protein is active in DNA replication in vitro. Virology 101:307–310

Arrand JE (1978) Mapping of adenovirus type 5 temperature-sensitive mutations by marker rescue in enhanced double DNA infections. J Gen Virol 41:573–586

Axelrod N (1978) Phosphoproteins of adenovirus 2. Virology 87:366–383

Babich A, Nevins JR (1981) The stability of early adenovirus mRNA is controlled by the viral 72 kd DNA-binding protein. Cell 26:371–379

Babiss LE, Ginsberg HS, Fisher PB (1983a) Cold-sensitive expression of transformation by a host range mutant of type 5 adenovirus. Proc Natl Acad Sci USA 80:1352–1356

Babiss LE, Young CSH, Fisher PB, Ginsberg HS (1983b) Expression of adenovirus E1a and E1b gene products and the Escherichia coli XGPRT gene in KB cells. J Virol 46:454–465

Babiss LE, Fisher PB, Ginsberg HS (1984) Deletion and insertion mutations in early region 1A of type 5 adenovirus producing cold-sensitive or defective phenotypes for transformation. J Virol 49:731–740

Begin M, Weber J (1975) Genetic analysis of adenovirus type 2. I. Isolation and genetic characterization of temperature-sensitive mutants. J Virol 15:1–7

Benzer S (1961) On the topography of the genetic fine structure. Proc Natl Acad Sci USA 47:403–415

Berget SM, Sharp PA (1979) Structure of late adenovirus 2 heterogeneous nuclear RNA. J Mol Biol 129:547–565

Berget SM, Flint SJ, Williams JF, Sharp PA (1976) Adenovirus transcription IV. Synthesis of viral-specific RNA in human cells infected with temperature-sensitive mutants of adenovirus 5. J Virol 19:879–889

Berget SM, Moore C, Sharp PA (1977) Spliced segments at the 5' terminus of adenovirus 2 late mRNA. Proc Natl Acad Sci USA 74:3171–3175

Berk AJ, Sharp PA (1978) Structure of the Adenovirus 2 early mRNAs. Cell 14:695–711

Berk AJ, Lee F, Harrison T, Williams J, Sharp PA (1979) Pre-early adenovirus 5 gene product regulates synthesis of early viral messenger RNAs. Cell 17:935–944

Berkner KL, Sharp PA (1983) Generation of adenovirus by transfection of plasmid DNA. Nucleic Acids Res 11:6003–6020

Bernards RA, Houweling A, Schrier P, Bos JL, Van Der Eb AJ (1982) Characterization of cells transformed by Ad5/Ad12 hybrid early region 1 plasmids. Virology 12D:422–432

Bhatti AR, Weber J (1979) Protease of adenovirus type 2: partial characterization. Virology 96:478–485

Bos JL, Polder LJ, Bernards R, Schrier PI, van den Elsen PJ, van der Eb AJ, van Ormondt H (1981) The 2.2 kb E1B mRNA of human Ad12 and Ad5 codes for two tumor antigens starting at different AUG triplets. Cell 27:121–131

Carlock LR, Jones NC (1981) Transformation-defective mutant of adenovirus type 5 containing a single altered E1a mRNA species. J Virol 40:657–664

Carstens EB, Magnan J, Weber J (1979) A dominant temperature-sensitive assembly mutant of adenovirus 2. Can J Microbiol 25:646–649

Carter TH, Blanton RA (1978a) Possible role of the 72,000-dalton DNA binding protein in regulation of adenovirus type 5 early gene expression. J Virol 25:664–674

Carter TH, Blanton RA (1978b) Autoregulation of adenovirus type 5 early gene expression. II. Effect of temperature-sensitive early mutations on virus RNA accumulation. J Virol 28:450–456

Carter TH, Ginsberg HS (1976) Viral transcription in KB cells infected by temperature-sensitive "early" mutants of adenovirus type 5. J Virol 18:156–166

Carter TH, Nicolas J-C, Young CSH, Fisher PB (1982) Multiple transformation phenotypes among revertants of temperature-sensitive mutants in the type 5 adenovirus DNA-binding protein. Virology 117:519–521

Cervera M, Dreffuss G, Penman S (1981) Messenger RNA is translated when associated with the cytoskeletal framework in normal and VSV-infected HeLa cells. Cell 23:113–120

Challberg MD, Kelly TJ Jr (1979) Adenovirus DNA replication in vitro. Proc Natl Acad Sci USA 76:655–659

Challberg MD, Kelly TJ Jr (1981) Processing of the adenovirus terminal protein. J Virol 38:272–277

Challberg SS, Ketner G (1981) Deletion mutants of adenovirus 2: Isolation and initial characterization of virus carrying mutations near the right end of the viral genome. Virology 114:196–209

Challberg MD, Desiderio SV, Kelly TJ Jr (1980) Adenovirus DNA replication in vitro: characterization of a protein covalently linked to nascent DNA strands. Proc Natl Acad Sci USA 77:5105–5109

Challberg MD, Ostrove JM, Kelly TJ Jr (1982) Initiation of adenovirus DNA replication: detection of covalent complexes between nucleotide and the 80-kilodalton terminal protein. J Virol 41:265–270

Chee-Sheung CC, Ginsberg HS (1982) Characterization of a temperature-sensitive fiber mutant of type 5 adenovirus and effect of the mutation on virion assembly. J Virol 42:932–950

Chinnadurai G (1983) Adenovirus 2 lp+ locus codes for a 19kd tumor antigen that plays an essential role in cell transformation. Cell 33:759–766

Chinnadurai G, Chinnadurai S, Brusca J (1979) Physical mapping of a large plaque mutation of adenovirus type 2. J Virol 32:623–628

Chow LT, Gelinas RE, Broker TR, Roberts RJ (1977) An amazing sequence arrangement at the 5′ ends of adenovirus 2 messenger RNA. Cell 12:1–8

Chow LT, Broker T, Lewis JB (1979a) The complex splicing patterns of RNA from the early regions of Ad2. J Mol Biol 134:265–303

Chow LT, Roberts JM, Lewis JB, Broker TR (1979b) A map of cytoplasmic RNA transcripts from lytic adenovirus type 2, determined by electron microscopy of RNA:DNA hybrids. Cell 11:819–836

Colby WW, Shenk T (1981) Adenovirus type 5 virions can be assembled in vivo in the absence of detectable polypeptide IX. J Virol 39:977–980

Cole CN, Crawford LV, Berg P (1979) Simian virus 40 mutants with deletions at the 3′ end of the early region are defective in adenovirus helper function. J Virol 30:683–691

Daniell E (1976) Genome structure of incomplete particles of adenovirus J Virol 19:685–708

D'Halluin JC, Milleville M, Boulanger PA, Martin GR (1978) Temperature-sensitive mutant of adenovirus type 2 blocked in virion assembly: accumulation of light intermediate particles. J Virol 26:344–356

D'Halluin JC, Milleville M, Martin GR, Boulanger P (1980) Morphogenesis of human adenovirus type 2 studied with fiber and penton base defective temperature-sensitive mutants. J Virol 33:88–99

D'Halluin J-C, Cousin C, Boulanger P (1982) Physical mapping of adenovirus type 2 temperature-sensitive mutations by restriction endonuclease analysis of interserotypic recombinants. J Virol 41:401–413

Downey JF, Rowe DT, Bacchetti S, Graham FL, Bayley ST (1983) Mapping of a 14,000 dalton antigen to early region 4 of the human adenovirus 5 genome. J Virol 45:514–523

Dunn AR, Hassell JA (1977) A novel method to map transcripts: Evidence for homology between an adenovirus mRNA and discrete, multiple regions of the viral genome. Cell 12:23–36

Edvardsson B, Everitt E, Jornvall H, Prage L, Philipson L (1976) Intermediates in adenovirus assembly. J Virol 19:533–547

Edvardsson B, Ustacelebi S, Williams J, Philipson L (1978) Assembly intermediates among adenovirus type 5 temperature-sensitive mutants. J Virol 25:641–651

Enomoto T, Lichy JH, Ikeda JE, Hurwitz J (1981) Adenovirus DNA replication in vitro: purification of the terminal protein in a functional form. Proc Natl Acad Sci USA 78:6779–6783

Ensinger MJ, Ginsberg HS (1972) Selection and preliminary characterization of temperature-sensitive mutants of type 5 adenovirus. J Virol 10:328–339

Epstein RH, Bolle A, Steinberg CM, Kellenberger E, Roy de la Tour E, Chevalley R, Edgar RS, Slisman M, Denhardt GH, Liesausis A (1963) Physiological studies of conditional lethal mutants of bacteriophage T4D. Cold Spring Harbor Symp Quant Biol 28:375–394

Eron L, Westphal H, Khoury G (1975) Post-transcriptional restriction of human adenovirus expression in monkey cells. J Virol 15:1256–1261

Esche H, Mathews MB, Lewis JB (1980) Proteins and messenger RNAs of the transforming region of wild-type and mutant adenoviruses. J Mol Biol 142:399–417

Evans RM, Fraser N, Ziff E, Weber J, Wilson M, Darnell JE (1977) The initiation sites for RNA transcription in Ad2 DNA. Cell 12:733–739

Everitt E, Sundquist B, Pettersson U, Philipson L (1973) Structural proteins of adenoviruses. X. Isolation and topography of low molecular weight antigens from the virions of adenovirus type 2. Virology 52:130–147

Feldman LT, Nevins JR (1983) Localization of the adenovirus E1Aa protein, a positive acting transcriptional factor, in infected cells. Mol Cell Biol 3:829–838

Feldman LT, Imperiale MF, Nevins JR (1982) Activation of early adenovirus transcription by the herpesvirus immediate early gene: Evidence for a common cellular control factor. Proc Natl Acad Sci USA 79:4952–4956

Fitzgerald M, Shenk T (1981) The sequence 5′-AAUAAA-3′ forms part of the recognition site for polyadenylation of late SV40 mRNAs. Cell 24:251–260

Flint SJ, Gallimore PH, Sharp P (1975) Comparison of viral RNA sequences in adenovirus 2-transformed and lytically infected cells. J Mol Biol 96:47–68

Flint SJ, Sambrook J, Williams JF, Sharp PA (1976) Viral nucleic acid sequences in transformed cells. IV. A study of the sequences of adenovirus 5 DNA and RNA in four lines of adenovirus 5-transformed. Virology 72:456–470

Fowlkes DM, Shenk T (1980) Transcriptional control regions of the adenovirus VAI RNA gene. Cell 22:405–413

Fowlkes DM, Lord ST, Linne T, Pettersson U, Philipson L (1979) Interaction between the adenovirus DNA-binding protein and double-stranded DNA. J Mol Biol 132:163–180

Fraser NW, Nevins JR, Ziff E, Darnell JR (1979) The major late adenovirus type 2 transcription unit: termination is downstream from the last poly(A) site. J Mol Biol 129:643–656

Friefeld BR, Krevolin MD, Horwitz MS (1983) Effects of the adenovirus H5ts125 and H5ts107 DNA binding proteins on DNA replication in vitro. Virology 124:380–389

Frost E, Williams J (1978) Mapping temperature-sensitive and host-range mutants of adenovirus type 5 by marker rescue. Virology 91:39–50

Gallimore PH, Williams J (1982) An examination of adenovirus type 5 mutants for their ability to induce group C adenovirus tumor-specific transplantation antigenicity in rats. Virology 120:146–156

Gallimore PH, Sharp PA, Sambrook J (1974) Viral DNA in transformed cells. II. A study of the sequences of adenovirus 2 DNA in nine lines of transformed rat cells using specific fragments of the viral genome. J Mol Biol 89:49–72

Galos RS, Williams J, Binger MH and Flint SJ (1979) Location of additional early gene sequences in the adenoviral chromosome. Cell 17:945–956

Galos RS, Williams J, Shenk T, Jones N (1980) Physical location of host range mutations of adenovirus type 5; deletion and marker rescue mapping. Virology 104:510–513

Gambke C, Deppert W (1983) Specific complex of the late nonstructural 100,000-dalton protein with newly synthesised hexon in adenovirus type 2-infected cells. Virology 124:1–12

Gaynor RB, Berk AJ (1983) Cis-acting induction of adenovirus transcription. Cell 33:683–693

Gingeras TR, Sciaky D, Gelinas RE, Bing-Dong J, Yen CE, Kelly MM, Bullock PA, Parsons BL, O'Neill KE, Roberts RJ (1982) Nucleotide sequences from the adenovirus 2 genome. J Biol Chem 257:13475–13491

Ginsberg HS (1979) Adenovirus structural proteins. In: Fraenkel-Conrat H, Wagner RR (eds) Comprehensive virology, vol 13. Plenum, New York, pp 409–457

Ginsberg HS, Young CSH (1977) The Genetics of Adenoviruses. In: Fraenkel-Conrat H, Wagner RR (eds) Comprehensive virology, vol 9. Plenum, New York, pp 27–88

Ginsberg HS, Ensinger MJ, Kauffman RS, Mayer AJ, Lundholm U (1974) Cell transformation: a study of regulation with types 5 and 12 adenovirus temperature-sensitive mutants. Cold Spring Harbor Symp Quant Biol 39:419–426

Goldman ND, Howley P, Khoury G (1981) Functional interaction between the early viral proteins of simian virus 40 and adenovirus. Virology 109:303–313

Graham FL, Abrahams PJ, Mulder C, Heijneker HL, Warnaar SO, deVries FAJ, Fiers W, van der Eb AJ (1974) Studies on in vitro transformation by DNA and DNA fragments of human adenoviruses and simian virus 40. Cold Spring Harbor Symp Quant Biol 39:637–650

Graham FL, Smiley J, Russell WC, Nairn R (1977) Characterization of a human cell line transformed by DNA from human adenovirus type 5. J Gen Virol 36:59–72

Graham FL, Harrison T, Williams J (1978) Defective transforming capacity of adenovirus type 5 host range mutants. Virology 86:10–21

Grodzicker T, Williams J, Sharp P, Sambrook J (1974) Physical mapping of temperature-sensitive mutations of adenoviruses. Cold Spring Harbor Symp Quant Biol 39:439–446

Grodzicker T, Anderson C, Sambrook J, Mathews MB (1977) The physical locations of structural genes in adenovirus DNA. Virology 80:111–126

Guilfoyle R, Weinmann R (1981) The control region for adenovirus VA RNA transcription. Proc Natl Acad Sci USA 78:3378–3382

Hammarskjold M-L, Winberg G (1980) Encapsidation of adenovirus 16 DNA is directed by a small DNA sequence at the left end of the genome. Cell 20:787–795

Harrison TJ, Graham FL, Williams JF (1977) Host range mutants of adenovirus 5 defective for growth in HeLa cells. Virology 77:310–329

Hassell JA, Weber J (1978) Genetic analysis of adenovirus type 2, VIII. Physical locations of temperature-sensitive mutations. J Virol 28:671–678

Hearing P, Shenk T (1983a) The adenovirus type 5 E1A transcriptional control region contains a duplicated enhancer element. Cell 33:695–703

Hearing P, Shenk T (1983b) Functional analysis of the nucleotide sequence surrounding the cap site for adenovirus type 5 region E1a mRNAs. J Mol Biol 167:809–822

Herisse J, Galibert F (1981) Nucleotide sequence of the Eco R1 E fragment of the adenovirus 2 genome. Nucleic Acids Res 9:1229–1240

Herisse J, Courtois G, Galibert F (1980) Nucleotide sequence of the EcoR1 D fragment of the adenovirus 2 genome. Nucleic Acids Res 8:2173–2192

Herisse J, Rigolet M, Dupont de Dinechin S, Galibert G (1981) Nucleotide sequence of the adenovirus 2 DNA fragment encoding for the carboxylic region of the fiber protein and the entire E4 region. Nucleic Acids Res 9:4023–4043

Ho Y-S, Galos R, Williams J (1982) Isolation of type 5 adenovirus mutants with a cold-sensitive host range phenotype: genetic evidence of an adenovirus transformation maintenance function. Virology 122:109–124

Horwitz MS (1978) Temperature-sensitive replication of H5ts125 adenovirus DNA in vitro. Proc Natl Acad Sci USA 75:4291–4295

Horwitz MS, Scharff MD, Maizel JV (1969) Synthesis and assembly of adenovirus 2. I. Polypeptide synthesis, assembly of capsomeres, and morphogenesis of the virion. Virology 39:682–697

Houweling A, Van den Elsen PJ, Van der Eb AJ (1980) Partial transformation of primary rat cells by the leftmost 4.5% fragment of adenovirus 5 DNA. Virology 105:537–550

Hudziak MS, Laski FA, RajBhandary UL, Sharp PA, Capecchi MR (1982) Establishment of mammalian cell lines containing multiple nonsense mutations and functional suppressor tRNA genes. Cell 31:137–146

Ikeda J, Enomoto T, Horwitz J (1981) Replication of adenovirus DNA-protein complex with purified proteins. Proc Natl Acad Sci 78:884–888

Ishibashi M, Maizel JV (1974) The polypeptides of adenovirus. V. Young virions, structural intermediates between top components and mature virions. Virology 57:409–924

Jochemsen H, Daniels GSG, Hertoghs JJ, Schrier PI, van den Elsen PJ, van der Eb AJ (1982) Identification of adenovirus type 12 gene products involved in transformation and oncogenesis. Virology 122:15–28

Jones N, Shenk T (1978) Isolation of deletion and substitution mutants of adenovirus type 5. Cell 13:181–188

Jones N, Shenk T (1979a) Isolation of adenovirus type 5 host range deletion mutants defective for translation of rat embryo cells. Cell 17:683–689

Jones NC, Shenk T (1979b) An adenovirus type 5 early gene function regulates expression of other early viral genes. Proc Natl Acad Sci USA 76:3665–3669

Kaplan LM, Ariga H, Hurwitz J, Horwitz MS (1979) Complementation of the temperature-sensitive defect in H5ts125 adenovirus DNA replication in vitro. Proc Natl Acad Sci USA 76:5534–5538

Katze MG, Persson H, Philipson L (1981) Control of adenovirus early gene expression: post-transcriptional control mediated by both viral and cellular gene products. Mol Cell Biol 1:807–813

Katze MG, Persson H, Philipson L (1982) A novel mRNA and a low molecular weight polypeptide encoded in the transforming region of adenovirus DNA. EMBO J 1:783–789

Kauffman RS, Ginsberg HS (1976) Characterization of a temperature-sensitive, hexon transport mutant of type 5 adenovirus. J Virol 19:643–658

Kelly TJ Jr, Lewis AM (1973) Use of non-defective adenovirus-simian virus 40 hybrids for mapping the simian virus 40 genome. J Virol 12:643–652

King J, Leak EV, Botstein D (1973) Mechanism of head assembly and DNA encapsulation in Salmonella phage P22. II. Morphogenetic pathway. J Mol Biol 80:697–731

Kitchingman GA, Lai S-P, Westphal H (1977) Loop structure in hybrids of early RNA and the separated strands of adenovirus DNA. Proc Natl Acad Sci USA 74:4392–4395

Klein H, Maltzman W, Levine AJ (1979) Structure – function relationship of the adenovirus DNA binding protein. J Biol Chem 254:11051–11060

Klessig D (1977) Two adenovirus mRNAs have a common 5′ terminal leader sequence encoded at least 10kb upstream from their main coding regions. Cell 12:9–21

Klessig DF, Anderson CW (1975) Block to multiplication of adenovirus serotype 2 in monkey cells. J Virol 16:1650–1668

Klessig DF, Chow LT (1980) Incomplete splicing and deficient accumulation of the fiber messenger RNA in monkey cells infected by human adenovirus type 2. J Mol Biol 139:221–242

Klessig DF, Grodzicker T (1979) Mutations that allow human Ad2 and Ad5 to express late genes in monkey cells map in the viral gene encoding the 72K DNA binding protein. Cell 17:957–966

Klessig DF, Quinlan MP (1982) Genetic evidence for separate functional domains on the human adenovirus specified 72kd, DNA binding protein. J Mol Appl Genet 1:263–272

Kruijer W, van Schik FMA, Sussenbach JS (1980) Nucleotide sequence analysis of a region of adenovirus 5 DNA encoding a hitherto unidentified gene. Nucleic Acids Res 8:6033–6042

Kruijer W, van Schaik FMA, Sussenbach JS (1981) Structure and organization of the gene coding for the DNA binding protein of adenovirus type 5. Nucleic Acids Res 9:4439–4457

Kruijer W, van Schaik FMA, Sussenbach JS (1982) Nucleotide sequence of the gene encoding adenovirus type 2 DNA binding protein. Nucleic Acids Res 10:4493–4500

Kruijer W, Nicolas J-C, van Schaik FMA, Sussenbach JS (1983) Structure and function of DNA binding proteins from revertants of adenovirus type 5 mutants with a temperature-sensitive DNA replication. Virology 124:425–433

Lai Fatt RB, Mak S (1982) Mapping of an adenovirus function involved in the inhibition of DNA degradation. Virology 42:969–977

Land H, Parada LF, Weinberg RA (1983) Tumorigenic conversion of primary embryo fibroblasts requires at least two cooperating oncogenes. Nature 304:596–602

Lassam NJ, Bayley ST, Graham FL (1978) Synthesis of DNA, late polypeptides and infectious virus by host-range mutants of adenovirus 5 in nonpermissive cells. Virology 87:463–467

Lassam NJ, Bayley ST, Graham FL (1979) Tumor antigens of human Ad5 in transformed cells and in cells infected with transformation-defective host-range mutants. Cell 18:781–791

Lebowitz J, Horwitz MS (1975) Synthesis and assembly of adenovirus polypeptides. III. Reversible inhibition of hexon assembly in adenovirus type 5 temperature-sensitive mutants. Virology 66:10–24

Levine AJ, van der Vliet PC, Rosenwirth B, Rabek J, Frenkel G, Ensinger M (1974) Adenovirus-infected, cell-specific DNA binding proteins. Cold Spring Harbor Symp Quant Biol 39:559–566

Lewis JB, Anderson CW (1983) Proteins encoded near the adenovirus late messenger RNA leader segments. Virology 127:112–123

Lewis JB, Mathews MB (1980) Control of adenovirus early gene expression: a class of immediate early products. Cell 21:303–313

Lewis AM, Levine AJ, Crumpacker CS, Levin MJ, Samabra RJ, Henry PH (1973) Studies of undefective adenovirus 2-SV40 hybrid viruses. V. Isolation of additional hybrids which differ in their SV40-specific biological properties. J Virol 11:655

Lewis JB, Atkins JF, Anderson CW, Baum PR, Gesteland RF (1975) Mapping of late adenovirus genes by cell-free translation of RNA selected by hybridization to specific DNA fragments. Proc Natl Acad Sci USA 12:1344–1348

Lewis JB, Anderson CW, Atkins JF (1977) Further mapping of late adenovirus genes by cell-free translation of RNA selected by hybridization to specific DNA fragments. Cell 12:37–44

Lichy JH, Horwitz MS, Hurwitz J (1981) Formation of a covalent complex between the 80,000 dalton adenovirus terminal protein and 5′ CMP in vitro. Proc Natl Acad Sci USA 78:2678–2682

Lichy JH, Field J, Horwitz MS, Hurwitz J (1982) Separation of the adenovirus terminal protein precursor from its associated DNA polymerase: role of both proteins in the initiation of adenovirus DNA replication. Proc Natl Acad Sci USA 79:5225–5229

Linne T, Philipson L (1980) Further characterization of the phosphate moiety of the adenovirus type 2 DNA-binding protein. Eur J Biochem 103:259–270

Logan J, Nicolas J-C, Topp WC, Girard M, Shenk T, Levine AJ (1981) Transformation by adenovirus early region 2A temperature-sensitive mutants and their revertants. Virology 115:419–422

Maat J, Van Ormondt H (1979) The nucleotide sequence of the transforming HindIII-G fragment of Ad5: the region between map positions 4.5 (HpaI site) and 8.0 (HindIII site). Gene 6:75–90

Mak I, Mak S (1983) Transformation of rat cells by cyt mutants of adenovirus type 12 and mutants of adenovirus type 5. J Virol 45:1107–1117

Martin GR, Warocquier R, Cousin C, D'Halluin JC, Boulanger PA (1978) Isolation and phenotypic characterization of human adenovirus type 2 temperature-sensitive mutants. J Gen Virol 41:303–314

Mathews M (1975) Genes for VA-RNA in Ad2. Cell 6:223–229

Mathews MB, Grodzicker T (1981) Virus-associated RNAs of naturally occurring strains and variants of group C adenoviruses. J Virol 38:849–862

Matsuo T, Wold WSM, Hashimoto S, Rankin A, Symington J, Green M (1982) Polypeptides encoded by transforming region E1b of human adenovirus 2: immunoprecipitation from transformed and infected cells and cell-free translation of E1b-specific mRNA. Virology 118:456–465

Mautner V, Williams J, Sambrook J, Sharp P, Grodzicker T (1975) The location of the genes coding for hexon and fiber proteins in adenovirus DNA. Cell 5:93–99

Maxam AM, Gilbert W (1977) A new method for sequencing DNA. Proc Natl Acad Sci USA 74:560–564

Mayer AJ, Ginsberg HS (1977) Persistence of type 5 adenovirus DNA in cells transformed by a temperature-sensitive mutant H5ts125. Proc Natl Acad Sci USA 74:785–788

McGrogan M, Raskas HJ (1978) Two regions of the adenovirus 2 genome specify families of late polysomal RNAs containing common sequences. Proc Natl Acad Sci USA 75:625–629

McKinnon RD, Bacchetti S, Graham FL (1982) Tn5 mutagenesis of the transforming genes of human adenovirus type 5. Gene 19:33–42

Mirza AM, Weber J (1977) Genetic analysis of adenovirus type 2. VII Cleavage modified affinity for DNA of internal virion proteins. Virology 80:83–97

Montell C, Fisher EF, Caruthers MH, Berk AJ (1982) Resolving the functions of overlapping viral genes by site-specific mutagenesis at a mRNA splice site. Nature 295:380–384

Montell C, Fisher EF, Caruthers MH, Berk AJ (1983) Inhibition of RNA cleavage but not polyadenylation by a point mutation in mRNA 3′ consensus sequence. Nature 305:600–605

Nevins JR (1981) Mechanism of activation of early viral transcription by the adenovirus E1A gene product. Cell 26:213–220

Nevins JR (1982) Induction of the synthesis of a 70,000 dalton mammalian heat shock protein by the adenovirus E1A gene product. Cell 29:913–919

Nevins JR, Darnell JE (1978) Groups of adenovirus type 2 mRNAs derived from a large primary transcript: probably nuclear origin and possible common 3′ ends. J Virol 25:811–823

Nevins JR, Wilson MC (1981) Regulation of adenovirus 2 gene expression at the level of transcriptional termination and RNA processing. Nature 290:113–118

Nevins JR, Winkler JJ (1980) Regulation of early adenovirus transcription: a protein product of early region 2 specifically represses region 4 transcription. Proc Natl Acad Sci USA 77:1893–1897

Nicolas J-C, Suarez F, Levine AJ, Girard M (1981) Temperature-independent revertants of adenovirus H5ts125 and H5ts107 mutants in the DNA binding protein: isolation of a new class of host-range temperature conditional revertants. Virology 108:521–524

Nicolas J-C, Ingrand D, Sarnow P, Levine AJ (1982) A mutation in the adenovirus type 5 DNA binding protein that fails to autoregulate the production of the DNA binding protein. Virology 122:481–485

Osborne TS, Gaynor RB, Berk AJ (1982) The TATA homology and the mRNA 5′ untranslated sequence are not required for expression of essential adenovirus E1a functions. Cell 29:139–148

Oosterom-Dragon EA, Anderson CW (1983) Polypeptide structure and encoding location of the adenovirus serotype 2 late, nonstructural 33K protein. J Virol 45:251–263

Oosterom-Dragon EA, Ginsberg HS (1980) Purification and preliminary immunological characterization of the type 5 adenovirus, nonstructural 100,000-dalton protein. J Virol 33:1203–1207

Oosterom-Dragon EA, Ginsberg HS (1981) Characterization of two temperature-sensitive mutants of type 5 adenovirus with mutations in the 100,000-dalton protein gene. J Virol 40:491–500

Ostrove JM, Rosenfeld P, Williams J, Kelly TJ Jr (1983) In vitro complementation as an assay for purification of adenovirus DNA replication proteins. Proc Natl Acad Sci USA 80:935–939

Paterson BM, Roberts B, Kuff EL (1977) Structural gene identification and mapping by DNA-mRNA hybrid-arrested cell-free translation. Proc Natl Acad Sci USA 74:4370–4374

Persson H, Sigmas C, Philipson C (1979) Purification and characterization of an early glycoprotein from Ad2-infected cells. J Virol 29:938–948

Persson H, Monstein H-J, Akusjarvi G, Philipson L (1981) Adenovirus early gene products may control viral mRNA accumulation and translation in vivo. Cell 23:485–496

Persson H, Katze MG, Philipson L (1982) Purification of a native membrane-associated tumor antigen. J Virol 42:905–917

Pettersson U, Philipson L (1975) Location of sequences on the adenovirus genome coding for the 5.5S RNA. Cell 6:1–4

Philipson L (1979) Adenovirus proteins and their messenger RNAs. Adv Virus Res 25:357–405

Philipson L, Pettersson U, Lindberg U, Tibbetts C, Vennstrom B, Persson T (1974) RNA synthesis and processing in adenovirus-infected cells. Cold Spring Harbor Symp Quant Biol 39:447–456

Proudfoot NJ, Brownlee GG (1974) Sequence at the 3′ end of globin mRNA shows homology with immunoglobin light chain in RNA. Nature 252:359–362

Rabek JP, Zakian VA, Levine AJ (1981) The SV40 A gene product suppresses the adenovirus H5ts125 defect in DNA replication. Virology 109:290–302

Rajagopalan S, Chinnadurai G (1981) Viable variants in VA-RNA I gene of an Ad2-Ad5 recombinant. Virology 112:564–571

Rassoulzadegan M, Cowie A, Carr A, Glaichenhaus N, Kamen R, Cuzin F (1982) The roles of individual polyoma virus early proteins in oncogenic transformation. Nature 300:713–718

Reich PR, Rose J, Forget B, Weissman SM (1966) RNA of low molecular weight in KB cells infected with Ad2. J Mol Biol 17:428–439

Ricciardi RP, Jones RL, Cepko CL, Sharp PA, Roberts BE (1981) Expression of early adenovirus genes requires a viral encoded acidic polypeptide. Proc Natl Acad Sci USA 78:6121–6125

Ross SR, Flint SJ, Levine AJ (1980a) Identification of the adenovirus early proteins and their genomic map positions. Virology 100:419–432

Ross SR, Levine AJ, Galos RS, Williams J, Shenk T (1980b) Early viral proteins in HeLa cells infected with adenovirus type 5 host-range mutants. Virology 103:475–492

Rowe DT, Graham FL (1983) Transformation of rodent cells by DNA extracted from transformation-defective adenovirus mutants. J Virol 46:1039–1044

Ruben M, Bacchetti S, Graham FL (1982) Integration and expression of viral DNA in cells transformed by host-range mutants of adenovirus type 5. J Virol 41:674–685

Ruley HE (1983) Adenovirus early region 1A enables viral and cellular transforming genes to transform primary cells in culture. Nature 304:602–606

Russell WC, Skehel JJ (1972) The polypeptides of adenovirus-infected cells. J Gen Virol 15:45–57

Russell WC, Newman C, Williams JF (1972) Characterization of temperature-sensitive mutants of adenovirus type 5-serology. J Gen Virol 17:265–279

Russell WC, Skehel JJ, Williams JF (1974) Characterization of temperature-sensitive mutants of adenovirus type 5: synthesis of polypeptides in infected cells. J Gen Virol 24:247–259

Sambrook J, Botchan M, Gallimore P, Ozanne B, Pettersson U, Williams J, Sharp PA (1974) Viral DNA sequences in cells transformed by simian virus 40, adenovirus type 2 and adenovirus type 5. Cold Spring Harbor Symp Quant Biol 39:615–632

Sambrook J, Williams J, Sharp PA, Grodzicker T (1975) Physical mapping of temperature-sensitive mutations of adenoviruses. J Mol Biol 97:369–390

Sanger F, Nicklen S, Coulson AR (1977) DNA sequencing with chain-terminating inhibitors. Proc Natl Acad Sci USA 74:5463–5467

Sarnow P, Hearing P, Anderson C, Reich N, Levine AJ (1982a) Identification and characterization of an immunologically conserved adenovirus early region E4-11K protein and its association with the nuclear matrix. J Mol Biol 162:565–583

Sarnow P, Ho YS, Williams J, Levine AJ (1982b) Adenovirus E1b-58kd tumor antigen and SV40 large tumor antigen are physically associated with the same 54kd cellular protein in transformed cells. Cell 28:387–394

Sarnow P, Hearing P, Anderson CW, Halbert DN, Shenk T, Levine AJ (to be published) The adenovirus E1B-58kd tumor antigen is physically associated with an E4-25kd protein in productively infected cells: identification and mapping of a new E4 region polypeptide.

Schrier PI, van den Elsen PJ, Hertoghs JJ, van der Eb AJ (1979) Characterization of tumor antigens in cells transformed by fragments of adenovirus type 5 DNA. Virology 99:372–385

Sharp PA, Gallimore PH, Flint SJ (1974) Mapping of adenovirus 2 RNA sequences in lytically infected cells and transformed cell lines. Cold Spring Harbor Symp Quant Biol 39:457–474

Shaw AR, Ziff EB (1982) Selective inhibition of Adenovirus type 2 early region II and III transcription by an anisomycin block of protein synthesis. Mol Cell Biol 2:789–799

Shenk T, Jones N, Colby W, Fowlkes D (1979) Functional analysis of adenovirus 5 host-range deletion mutants defective for transformation of rat embryo cells. Cold Spring Harbor Symp Quant Biol 44:367–375

Shiroki K, Maruyama K, Saito I, Fukui Y, Shimojo H (1981) Incomplete transformation of rat cells by a deletion mutant of adenovirus type 5. J Virol 38:1048–1054

Shortle D, DiMaio D, Nathans D (1981) Directed mutagenesis. Annu Rev Genet 15:265–294

Signas C, Katze MG, Persson H, Philipson L (1982) An adenovirus glycoprotein binds heavy chains of class I transplantation antigens from man and mouse. Nature 299:175–178

Smart JE, Stillman BW (1982) Adenovirus terminal protein precursor: partial amino acid sequence and the site of covalent linkage to virus DNA. J Biol Chem 257:13499–13506

Smart JE, Lewis JB, Mathews MB, Harter ML, Anderson CW (1981) Adenovirus type 2 early proteins: Assignment of the early region 1A proteins synthesized in vivo and in vitro to specific mRNAs. Virol 112:703–713

Solnick D (1981) An adenovirus mutant defective in splicing RNA from early region 1A. Nature 291:508–510

Solnick D (1983) Shuffling adenovirus promoters: a viral recombinant with early region 1A under late transcriptional control. EMBO 2:845–851

Solnick D, Anderson MA (1982) Transformation-deficient adenovirus mutant defective in expression of region E1a but not region E1b. J Virol 42:106–113

Stillman BW (1983) The replication of adenovirus DNA. UCLA Symp Mol Cell Biol 10:381–393

Stillman BW, Tamanoi F (1983) Adenovirus DNA replication: DNA sequences and enzymes required for initiation in vitro. Cold Spring Harbor Symp Quant Biol 47:741–752

Stillman BW, Lewis JB, Chow LT, Mathews MB, Smart JE (1981). Identification of the gene and mRNA for the adenovirus terminal protein precursor. Cell 23:497–508

Stillman BW, Tamanoi F, Mathews MB (1982) Purification of an adenovirus-coded DNA polymerase that is required for initiation of DNA replication. Cell 31:613–623

Stinsky MF, Ginsberg HS (1974) Antibody to the type 5 adenovirus hexon polypeptide: detection of nascent polypeptides in cytoplasm of infected KB cells. Intervirology 4:226–236

Stow ND (1981) Cloning of a DNA fragment from the left-hand terminus of the adenovirus type 2 genome and its use in site-directed mutagenesis. J Virol 37:171–180

Takemori N, Riggs JL, Aldrich C (1968) Genetic studies with tumorigenic adenoviruses. I. Isolation of cytocidal (cyt) mutants of adenovirus type 12. Virology 36:575–586

Tamanoi F, Stillman BW (1982) Function of adenovirus terminal protein in the initiation of DNA replication. Proc Natl Acad Sci USA 79:2221–2225

Tarodi B, Blair GE, Rekosh DMK, Russell WC (1979) Characterization of two temperature-sensitive mutants of adenovirus type 5. J Gen Virol 43:531–540

Thimmappaya B, Jones N, Shenk T (1979) A mutation which alters initiation of transcription by RNA polymerase III on the Ad5 chromosome. Cell 18:947–954

Thimmappaya B, Weinberger C, Schneider RJ, Shenk T (1982) Adenovirus VAI RNA is required for efficient translation of viral mRNAs at late times after infection. Cell 31:543–551

Thomas GP, Mathews MB (1980) DNA replication and the early to late transition in adenovirus infection. Cell 22:523–533

Thummel C, Tjian R, Hu S-L, Grodzicker T (1983) Translational control of SV40 T antigen expressed from the adenovirus late promoter. Cell 33:455–464

Tibbetts C (1977) Viral DNA sequences from incomplete particles of human adenovirus type 7. Cell 12:243–249

Tigges MA, Raskas HJ (1982) Expression of adenovirus-2 early region 4: assignment of the early region 4 polypeptides to their respective mRNAs using in vitro translation. J Virol 44:907–921

Tooze J (1980) DNA Tumor viruses. Cold Spring Harbor Laboratory, New York

Ustacelebi S, Williams JF (1972) Temperature-sensitive mutants of adenovirus defective in interferon induction at non-permissive temperature. Nature 235:52–53

van der Eb AJ, Mulder C, Graham FL, Houweling A (1977) Transformation with specific fragments of adenovirus DNAs. 1. Isolation of specific fragments with transforming activity of adenovirus 2 and 5 DNA. Gene 2:115–132

van der Eb AJ, van Ormondt H, Schrier PI, Lupker JH, Jochemsen H, van den Elsen H, DeLeys PJ, Maat J, van Bevern CP, Dijkema R, DeWard A (1979) Structure and function of the transforming genes of human adenovirus and SV40. Cold Spring Harbor Symp Quant Biol 44:383–399

Van der Vliet PC, Levine AJ (1973) DNA binding proteins specific for cells infected by adenovirus. Nature New Biol 239:47–49

Van der Vliet PC, Sussenbach JS (1975) An adenovirus type 5 gene function required for initiation of viral DNA replication. Virology 67:415–426

Van der Vliet PC, Levine AJ, Ensinger MJ, Ginsberg HS (1975) Thermolabile DNA binding proteins from cells infected with a temperature-sensitive mutant of adenovirus defective in viral DNA synthesis. J Virol 15:348–354

Van der Vliet PC, Blanken WM, Zandberg J, Jansz HS (1977) Function of an early DNA binding protein in the replication of adenovirus DNA. In: Molineux I, Kolinyama M (eds) DNA synthesis, present and future. Plenum, New York, pp 869–885

Van der Vliet PC, Keegstra W, Jansz HS (1978) Complex formation between the adenovirus type 5 DNA binding protein and single-stranded DNA. Eur J Biochem 86:389–398

Velicer LF, Ginsberg HS (1970) Synthesis, transport, and morphogenesis of type 5 adenovirus capsid proteins. J Virol 5:338–352

Volkert FC, Young CSH (1983) The genetic analysis of recombination using adenovirus overlapping terminal DNA fragments. Virology 125:175–193

Weber J (1976) Genetic analysis of adenovirus type 2. III. Temperature sensitivity of processing of viral proteins. J Virol 17:462–471

Weinberg DH, Ketner G (1983) A cell line that supports the growth of a defective early region 4 deletion mutant of human adenovirus type 2. Proc Natl Acad Sci USA 80:5383–5386

Weinmann R, Raskas HJ, Roeder RG (1974) Role of DNA-dependent RNA polymerases II and III in transcription of the adenovirus genome late in productive infection. Proc Natl Acad Sci USA 71:3426–3430

Wilkie NM, Ustacelebi S, Williams JF (1973) Characterization of temperature-sensitive mutants of adenovirus type 5: nucleic acid synthesis. Virology 51:499–503

Williams JF, Ustacelebi S (1971) Complementation and recombination with temperature-sensitive mutants of adenovirus type 5. J Gen Virol 13:345–348

Williams JF, Gharpure M, Ustacelebi S, McDonald S (1971) Isolation of temperature-sensitive mutants of adenovirus type 5. J Gen Virol 11:95–101

Williams JF, Young CSH, Austin PE (1974) Genetic analysis of human adenovirus type 5 in permissive and nonpermissive cells. Cold Spring Harbor Symp Quant Biol 39:427–437

Williams JF, Grodzicker T, Sharp P, Sambrook J (1975) Adenovirus recombination: physical mapping of crossover events. Cell 4:113–119

Williams JF, Galos RS, Binger MH, Flint SJ (1979) Location of additional early regions within the left quarter of the adenoviral genome. Cold Spring Harbor Symp Quant Biol 44:353–365

Yee S-P, Rowe DT, Tremblay ML, McDermott M, Branton PE (1983) Identification of human adenovirus early region 1 products by using antisera against synthetic peptides corresponding to the predicted carboxy termini. J Virol 46:1003–1013

Yeh-Kai L, Akusjarvi G, Alestrom P, Pettersson U, Tremblay M, Weber J (1983) Genetic identification of an endoproteinase encoded by the adenovirus genome. J Mol Biol 167:217–222

Young CSH, Williams JF (1975) Heat-stable variant of human adenovirus type 5: characterization and use in three-factor crosses. J Virol 15:1168–1175

Young CSH, Shenk T, Ginsberg HS (1983) The genetic system. In: Fraenkel-Conrat H, Wagner RR (eds) The viruses: Adenovirus. Plenum, New York

Zain S, Sambrook J, Roberts RJ, Keller W, Fried M, Dunn A (1979) Nucleotide sequence analysis of the late leader segments in a cloned copy of adenovirus 2 fiber mRNA. Cell 16:851–861

Ziff E, Evans R (1978) Coincidence of the promoter and capped 5' terminus of RNA from the adenovirus 2 major late transcription unit. Cell 15:1463–1475

Ziff E, Fraser NW (1978) Adenovirus type 2 late mRNA: structural evidence for 3' coterminal species. J Virol 25:897–906

Adenovirus DNA Replication

J. FÜTTERER and E.-L. WINNACKER

1 Introduction

The molecular biology of adenoviruses is understood in considerable detail. This relates not only to the primary structure of the viral genome, transcription patterns, and virus/host cell interactions, but also to the mechanism of viral DNA replication. This review will document that adenovirus DNA replication occurs via a novel mechanism for the initiation reaction. In addition, adenovirus replication also represents the only known system that can be initiated efficiently in vitro.

2 Structure of the Viral Genome

Adenoviruses are known to interact with a variety of vertebrate hosts. The existence of almost 100 different serotypes in species ranging from chicken to humans is well documented. The human adenovirus serotypes 1–38 have been subdivided into five groups on the basis of DNA homology (GREEN et al. 1979). The adenovirus serotypes 2 (Ad2) and 5 (Ad5), which are members of group C nononcogenic human adenoviruses, have been used extensively as model systems for the study of DNA replication. Chicken virus CELO, mouse adenovirus FL (AdFl), simian virus SA7, and other human serotypes, notably Ad7 (group B) and type 12 (group A) have also been studied occasionally. The genome of Ad2 is approximately 36000 bp in length. After the sequences for regions between coordinates 0–32% (GINGERAS et al. 1982; ALESTRÖM et al. 1982a), 70–100% (GALIBERT et al. 1979; HERISSE et al. 1980; HERISSE and GALIBERT

Institut für Biochemie der Universität München, Karlstr. 23, D-8000 München 2

```
AD2    CATCATC-ATAATATACCTTATT--TTGGATT-GAAGCCAATATGATAATGAGGGGGTGGAGTTTGTGACGTGGCGCGGGGCGTGG
AD5    CATCATCAATAATATACCTTATT--TTGGATT-GAAGCCAATATGATAATGAGGGGGTGGAGTTTGTGACGTGGCGCGGGGCGTGG
AD3    CTATCTATATAATATACCTTATAG-ATGGAAT-GGTGCCAACATGTAAATGAGGTAATTTAAAAAAGTGCGCGCTGTGTGGTGATT
AD7    CTCTCTATATAATATACCTTATAG-ATGGAAT-GGTGCCAACATGTAAATGAGGTAATTTAAAAAAGTGCGCGCTGTGTGGTGATT
AD4    CTCTCTCTATAATATACCTTATTTTTTTTGTG-TGAGTTAATATGGCAATAAGGCGTGAAAATTTGGGGATGGGGCGCGCTGATTG
AD9    CTATCTATATAATATACCCCACAAAGTAAACA-AAAGTTAATATGCAAATGAGCTTTTGAATTTTAACGGTTTCGGGGCGGAGCCA
AD10   CTTCATCAATAATATACCCCACAAAGTAAACA-AAAGTTAATATGCAAATGAGCTTTTGAATTTTAACGGTTTCGGGGCGGAGCCA
AD12   CATCATCAATAATATACCTTATA--CTGGACT-AGTGCCAATATTAAAATGAAGTGGGCGTAGTGTGTAATTTGATTGGGTGGAGG
AD18   CATCATCAATAATATACCTTATA--CTGGACT-AGAGCCAATATTAAAATGAAGTGGGTGTGGCGATGTACTTTGATTGGGTGGAG
AD31   CATCATCAATAATATACCTTACA--CTGGACT-TGAGCCAATATTAAAATGAAGTGGGCGGAGTGAATAGTTAATTGACCGTAGGC
SA7    CATCATCAATAATATACCTTATT--TTGGGAACGGTGCCAATATGCTAATGAGGTGGGCGGAGTTTGGTGACGTATGCGGAAATGG
TAV    CATCATCAATAATATACCTGACACTTTTGACGT-------------AATGACGGTTGCAAGTGCCACGTCGTCGTGGGCGTGTCT
ADFL   CATCATCAATAATATAC-T--A--------------GTTAGCA-AAAAATGGCGCCTTTGTTTGGCTTTGTTCCAACTGTTTTTGG
CELO   GATGATGTATAATA-ACCTCAAAA----------ACTAACGCAGTCATAACCGGCCATAACCGCAGCGTGTCGC - 63
```

```
AD2    GAACGGGGCGGGTGACGTAG - 102
AD5    GAACGGGGCGGGTGACGTAG - 103
AD3    GGCTGCGGGGTTAACGGCTAAAAGGGGCGGCGCGACCGTGGGAAAATGACGT - 136
AD7    GGCTGTGGGGTGAATGACTAACATGGGCGGGGCGGCCGTGGGAAAATGACGT - 136
AD4    GCTGTGACAGCGGCGTTCGTTAGGGGCGGGG - 116 - CAGGTGACGTTTTGAT / TAACTGATGTGTTTAA
AD9    ACGCTGATTGGACAGAGAAGACGATGCAAATGACGTCACGACGCACGGCTAACGGTCGCCGCGGAGGCGGGGC - 158
AD10   ACGCTGATTGGACAGAGAAGACGATGCAAATGACGTCACGACGCACGGC - 135
AD12   TGTGGCTTTGGCGTGCTTGTAAGTTTGGGCGGATGAGGAAGTGGGGCGCGGCGTGGGAGCCGGCGCGCCGGTGTGACGT - 162
AD18   GTGTGGCCTGGGCGTGTTTGTAAGTTTGGGCGGATGAGGAAGTTGGGCGGCGCGGCGTGGGAGCCGGCGCGCCGGTGTGACGTGT - 165
AD31   GTGGTTTGCAAGTTTGCCGAAGCCGGATAGTGACGCGTGTGGGAGCCGGGCGCGCGCCGGATGTGACG - 148
SA7    GCGGAGTTAGGGGCGGGGTTTGGCGGTAGGCGTGGCT........
TAV    TTTGTGACCTTTGGACGGGCGTTTCGCTGGCCGGGTTCCCAGTTTCGGGGCCGTTCCCGAGAACGTTGAGTCATGACAGCTGACCCGGG - 161
ADFL   CCCGAGTTGGGTTTCGTTTTCCCGGG - 93
```

Fig. 1. DNA sequences of adenovirus inverted repeats. The nucleotide sequences of known adenovirus ITRs are shown. The *numbers* correspond to the last nucleotide position determined at the 3′ positions. In the case of Ad4, sequences adjacent to the 3′ end at the right and left sides of the genome are also shown. Highly conserved sequences in the AT-rich part are indicated by *boxes*, sequences in the GC-rich part which are homologous to sequences at the origin of papovaviruses are underlined (*wavy line*). The sequence TGACGT, which occurs near the end of primate adenoviruses, is also indicated (*straight line*). Sequence data are taken from STEENBERGH et al. (1977) (Ad5), TOLUN et al. (1979) (Ad3, Ad7, Ad12, SA7), ARRAND and ROBERTS (1979) (Ad2), DIJKEMA and DEKKER (1979) (Ad3, Ad7), SHINAGAWA and PADMANABHAN (1980) (Ad2, Ad12), SUGISAKI et al. (1980) (Ad12), TEMPLE et al. (1981) (AdFl), GARON et al. (1982) (Ad18), STILLMAN et al. (1982a), (Ad4, Ad9, Ad10, Ad31), ALESTRÖM et al. (1982b) (CELO), and BRINCKMANN et al. (1983) (Tupaia adenovirus TAV)

1981), various other segments of special interest, e.g., the gene for the 72K DNA-binding protein (DBP) (KRUIJER et al. 1982), have been reported. Now the complete primary sequence of Ad2 is known (R.J. ROBERTS, personal communication).

There are two novel and unique features of the adenovirus genome that are related to the mechanism of viral DNA replication, i.e., the inverted terminal repetition (ITR) and the terminal protein (TP). The sequences from the ITRs from human adenoviruses, comprising members of all five homology groups, simian adenovirus SA7, mouse AdFl, Tupaja adenovirus (TAV), and CELO virus have been determined. All ITRs extend to the very end of the viral genome,

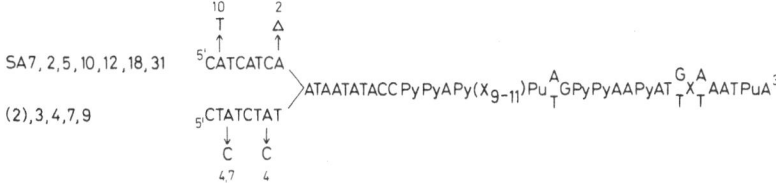

Fig. 2. Consensus sequence for the AT-rich region of the inverted terminal repetitions of primate adenovirus DNAs. The first eight nucleotides are related to either the Ad5 or the Ad3 sequence from which all other serotype sequences can be derived by base changes or deletions as indicated. In different plaque isolates of one serotype sequences of both types can be observed. The sequence from position 9 to position 18 is highly conserved in all known adenoviruses (see Fig. 1). (Redrawn from STILLMAN et al. 1982a)

the first nucleotide on the 5′ end always being dC except for CELO virus, where it is dG. The sizes of the ITRs range from 63 bp (CELO virus) to 165 bp (Ad12 and Ad18) (Fig. 1). The primary sequences show high degrees of homology both for viruses within one group and for all known adenoviruses. To facilitate sequence comparisons, it is useful to divide each ITR into an AT-rich region of 50–52 bp in length, positioned at the immediate termini of adenovirus DNA molecules, and a GC-rich region between 50 and 110 bp long, extending into the interior of the DNA molecules. In general the first eight base pairs can be of either of two types, which are characteristic for Ad5 (CATCATCA) or Ad3 (CTATCTAT). These sequences appear to be interchangeable, since we have sequenced the ITR of a plaque-purified (GRÖGER et al., unpublished work) authentic Ad2 molecule containing the terminus typical for Ad3. The terminal eight nucleotides are followed by a stretch of ten base pairs (ATAATA-TACC), which is perfectly conserved in all human serotypes and which is present also in the ITR of nonhuman adenoviruses (Fig. 2). Towards the interior part of the ITRs the various sequences diverge rapidly. Nevertheless, there are two sequence features in the GC-rich part of the molecule that are common to most or all sequences. The sequence GGGCGG occurs at least once within the ITRs and the sequence TGACGT is observed at or near the end of the ITRs (Fig. 1).

It is not known whether these and the previously noted sequence features are of any significance. It is most likely, however, that they represent the origin of adenovirus DNA replication.

The second unique structural element of adenovirus DNA is the terminal protein. Viral DNA can be isolated from viruses in a form which contains a 55K protein covalently linked to each 5′ terminus (ROBINSON et al. 1973; ROBINSON and BELLETT 1974; REKOSH et al. 1977). Linkage occurs via a phosphodiester bond between the β-OH of a particular serine residue (see below) in the protein and the 5′-OH group of the terminal deoxycytidine residue (DESIDERIO and KELLY 1981). The 55K protein is synthesized as an 80K precursor found at the 5′-ends of replicating DNA molecules in the nuclei of infected cells (COOMBS et al. 1979; KELLY and LECHNER 1979; STILLMAN and BELLETT 1979; VAN WIELINK et al. 1979; CHALLBERG et al. 1980). Late in the infection cycle the precursor is cleaved by a virus-coded protease to yield the mature

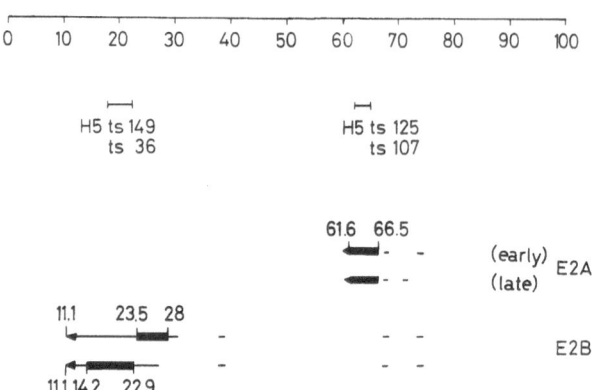

Fig. 3. Map of Ad2-encoded replication proteins. The regions of the respective mRNAs that code for the main bodies of the proteins are indicated by *thick lines*. The E2A region is transcribed from both early and late promoters and encodes for the DNA-binding 72K protein (for review see FLINT and BROKER 1980); the E2B region is transcribed from the early promoter (STILLMAN et al. 1981) and codes for pTP (positions 28–23.5) and a protein with DNA polymerase activity (positions 22.9–14.2), as described in the text. The map positions of Ad5 mutants that are temperature-sensitive for DNA replication are also shown

55K protein (CHALLBERG and KELLY 1981). A possible intermediate of 62K has been discussed (STILLMAN et al. 1981). This processing step only occurs after the virus particle has been fully assembled (FÜTTERER 1982).

The origin of the precursor of the terminal protein (pTP) had long been an enigma until the formation of an 80K protein could be demonstrated by cell-free translation of virus-specific mRNA selected by hybridization against the viral l-strand between coordinates 11 and 31.5 (STILLMAN et al. 1981). This 80K protein was shown to be structurally related both to the 55K TP and to the 80K protein which is covalently associated with replicating viral DNA. These observations led to the definition of a new early transcription unit, designated E2B (Fig. 3). Transcripts from this region originate from a promoter located at coordinate 75. They possess leader sequences at coordinates 75 and 39, which are spliced to three main bodies at coordinates 30.3, 26.0, or 23.1, and terminate at a common termination site around coordinate 11.1. The sequence of the l-strand of the viral DNA reveals two long open reading frames (GINGERAS et al. 1982; ALESTRÖM et al. 1982a). One of these begins with an ATG at nucleotide 10534 (28.97%) and ends with a TAG termination codon at nucleotide 8575 (23.48%). It encodes a 74.6K polypeptide composed of 668 aminoacids. A comparison of the methionine-containing tryptic peptides of the pTP with those predicted from this reading frame established the identity of both sequences (SMART and STILLMAN 1982). SMART and STILLMAN (1982) also studied and identified the tryptic peptides which contain the site binding to the 5′-ends of Ad2 DNA. The predicted binding site includes serine residue 577, encoded by nucleotides 8850–8852 (coordinate 24.2).

The 55K terminal protein derives from the COOH terminal end of pTP. The exact cleavage site is not known. A sequence within pTP (asp-met-thr-gly-gly-val-phe) around map coordinate 26.1, however, resembles the proteolytic cleavage site (asn-met-ser-gly-gly-ala-phe) utilized for maturation of protein pVI to virion protein VI (AKUSJÄRVI and PERSSON 1981) and thus might indeed represent the N-terminus of the terminal 55K protein.

The 80K dalton pTP functions in the initiation of DNA replication by covalently binding dCTP to form a pTP–dCMP complex (CHALLBERG et al. 1982) in an ATP-dependent reaction (DE JONG et al. 1983). This reaction is inhibited by antisera against the terminal protein (RIJNDERS et al. 1983a). The parental TP that is covalently attached to the template DNA is not required for this reaction. The initiation event rather depends on the presence of specific DNA sequences at the termini of a linear DNA molecule (TAMANOI and STILLMAN 1982; VAN BERGEN et al. 1983). It thus remains unknown whether there is a specific biological function to the 55K TP after it has been processed from its precursor. It is known for certain that the TP enhances the infectivity of adenovirus DNA in transfection experiments by a factor of 3–100 (SHARP et al. 1976; CHINNADURAI et al. 1978; ESTES 1978).

3 Viral DNA in Lytically Infected Cells

In human cells productively infected with Ad2 and Ad5 the onset of viral DNA replication usually occurs between 8 and 12 h after infection. The subsequent replication process, which takes place in the nucleus of the infected cells, reaches a maximum at 18–22 h after infection. While this time course of events applies to truly permissive human HeLa and A549 (derived from a human lung carcinoma) cells, for example, it can be delayed considerably in less permissive systems (e.g., ANTOINE et al. 1982). It is therefore useful and, as far as the study of viral DNA replication is concerned also important, to determine this parameter carefully for any particular adenovirus/host cell system. Several forms of the viral DNA have been detected in productively infected cells. Parental and newly synthesized viral DNA have been observed to exist in high-molecular-weight forms sedimenting at 50–100S in alkaline sucrose gradients as early as 2–5 h after infection. A structural analysis of these fast-sedimenting forms by equilibrium sedimentation, reassociation kinetics, and restriction enzyme analysis has convincingly demonstrated that they exist in a covalent linkage with cellular DNA sequences (SCHICK et al. 1976). It is also well established from pulse-chase experiments that they do not represent true precursors of mature viral DNA. In addition, studies with the adenovirus mutants H5ts36 and H5ts125, which are defective in viral DNA replication, clearly demonstrate that integration and replication are two independent events (TYNDALL et al. 1978). The significance of the high-molecular forms of the viral DNA for viral DNA replication thus remains unknown.

Recently, covalently closed circles of Ad5 DNA have been isolated from infected permissive and non-permissive cells (RUBEN et al. 1983). These structures amount to 8–15% of the intracellular viral DNA in (semipermissive) BHK cells infected with an hr-1 mutant of Ad5, and this proportion remains constant from 5 to 120 h after infection. The occurrence of the circles as early as 3–5 h after infection suggests that incoming DNA molecules may be circularized prior to the onset of viral DNA replication. On the other hand, circular DNA molecules cannot be extracted from virions and thus appear never or rarely to be packaged in vivo. It is therefore most likely that adenovirus DNA circles play a role in integration rather than in replication events.

An intriguing but unsolved problem concerns the structure of the adenovirus chromatin. Unlike other DNA viruses which replicate in the nucleus, such as the papovaviruses, adenoviruses do not use cellular histones to condense and package viral DNA into virions. Instead, the viral DNA is complexed with viral-coded proteins (PRAGE and PETTERSSON 1971; BROWN et al. 1975). The major core protein, protein VII, a small arginine-rich protein (39 of 173 residues; SUNG et al. 1983) with an N-terminal basic domain similar to that of histones, binds tightly to viral DNA (LISCHUE and SUNG 1977). It is synthesized in the form of a 22K precursor, pVII, which is processed to mature protein VII only after the pVII/DNA complex has entered the viral capsid. The synthesis of pVII and the other minor core proteins (polypeptides V and X) proceeds concomitantly with viral DNA synthesis at high rates. Cellular DNA and histone synthesis have already ceased at this stage of the infection cycle. It is thus unlikely that the adenovirus chromatin will be arranged in the same repeating units or nucleosomes that are typical for cellular chromatin. In fact, probing of intracellular adenovirus DNA at a late stage in the infection cycle by digestion with micrococcal nuclease, revealed only a diffused protected band, smaller than 160 bp, but no discrete multimers. Similar results are obtained after limited nuclease digestion of viral cores (BROWN and WEBER 1980; DANIELL et al. 1981). In contrast, viral chromatin early in infection exhibits a repeat pattern rather similar to the cellular nucleosomal repeat (SERGEANT et al. 1979; TATE and PHILIPSON 1979). It is possible that shortly after infection cellular histones replace the core proteins of incoming viral DNA as virions are uncoated. Late in infection, the newly synthesized viral DNA associates with the newly synthesized polypeptide pVII, yielding a structure with a different digestion pattern. It has been proposed that this structure consists of six molecules of protein VII and 200 bp of DNA (CORDEN et al. 1976). It is not clear at present why the late adenovirus chromatin shows no characteristic repeat pattern. SUNG et al. (1983) have identified a protamine-like domain in polypeptide pVII. If this interacts with adjacent nucleosome regions it could render the spaced DNA less susceptible to nuclease attack. Furthermore, there may be different pools of adenovirus DNA late in infection which all have different protein associations. Even if newly synthesized viral DNA exhibited a definite digestion pattern it would be obscured by the contribution of other structures. The literature abounds with assumptions on the existence of different types of viral DNA, integration complexes, transcription complexes, and replication complexes. Only careful in vitro reconstitution studies will eventually permit clarification.

4 Mechanism of Adenovirus DNA Replication

Intact mature forms of the viral DNA are observed in infected cells from 8 to 12 h after infection and may represent the largest proportion among the

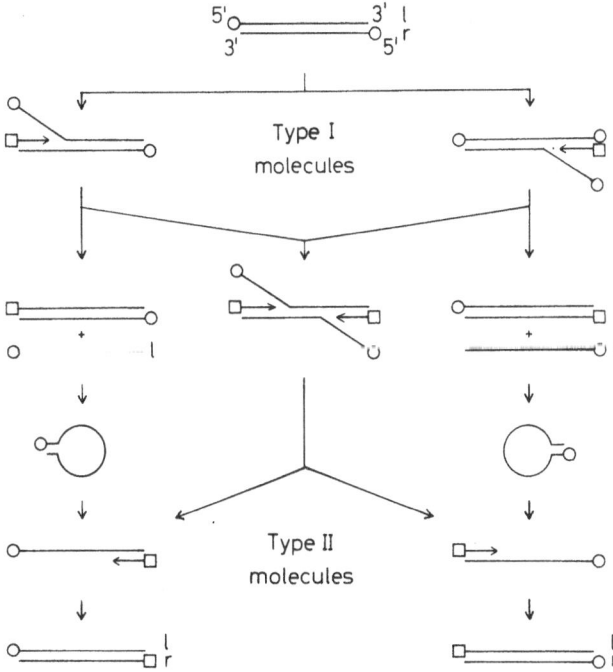

Fig. 4. Replication pattern of adenovirus DNA. As described in the text, DNA replication occurs in two distinct steps, by displacement and complementary strand synthesis. Type I and II molecules are the corresponding replicative intermediates of these processes. The initiation reaction of the displacement synthesis requires a pTP (□) and can occur at both molecular ends of the linear duplex DNA. The pan-handled structures of single-stranded DNA are purely hypothetical, but possible conformations of the viral DNA and serve to demonstrate the functional equivalence of the molecular termini. (Modified and redrawn from LECHNER and KELLY 1977)

various forms of viral DNA. Copy numbers in infected cell nuclei were found to increase to up to 100000 and more mature DNA molecules at late times after infection (FLINT et al. 1976; FANNING and DOERFLER 1977). Replicative forms of the viral DNA were originally recognized by pulse or pulse-chase experiments (PEARSON and HANAWALT 1971; VAN DER VLIET and SUSSENBACH 1972; VAN DER EB 1973). They are distinguished from mature viral DNA by an increased sedimentation rate in neutral sucrose gradients, an increase in buoyant density of 6–10 mg/ml in neutral cesium chloride, retention on BND-cellulose (ROBIN et al. 1973), and sensitivity to single-stranded DNA specific nucleases (PETTERSSON 1973; ROBIN et al. 1973). These parameters point to an extended single-stranded character of replicating adenovirus DNA. Careful hybridization analyses have shown that these single-stranded regions, which amount to 15–30% of the total pool of replicating DNA molecules, arise from all regions and from both complementary strands of the viral genome (PETTERSSON 1973; LAVELLE et al. 1975).

Electron microscopic studies have supported these conclusions and led to the recognition of two classes of replicating DNA molecules (ELLENS et al. 1974; LECHNER and KELLY 1977; REVET and BENICHOU 1981). Type I molecules (Fig. 4) are linear full-length duplices with at least one and up to four single-stranded branches. Type II molecules (Fig. 4) represent unbranched full-length DNA molecules with a single- and a double-stranded region. Both types amount to over 90% of all replicating DNA molecules, while only a small fraction contains features of both type I and type II molecules. As shown by partial denaturation mapping (LECHNER and KELLY 1977) and by denaturation following in vivo crosslinking of DNA strands (REVET and BENICHOU 1981), the growing points in either type I or type II molecules arise with equal frequency from both ends of the viral genome.

These observations are compatible with the following two-step model of adenovirus DNA synthesis (Fig. 4; LECHNER and KELLY 1977). In step one, replication starts at either molecular end of the viral DNA by synthesis to a daughter strand in $5' \rightarrow 3'$ direction with concomitant displacement of the parental strand of the same polarity. In the case of initiation at the molecular right-hand end, the displaced strand is the parental r-strand; in the case of initiation at the molecular left-hand end, it is the viral l-strand. Type I molecules, the intermediates of the first step in viral DNA synthesis, do not contain the typical replication forks known from Cairns-type replication intermediates. Thus, instead of synchronous replication of both parental strands, only one parental strand is replicated and only one daughter molecule is produced from any given displacement fork. Adenovirus DNA synthesis thus occurs in an asymmetric and semiconservative fashion.

A second daughter molecule is obtained by complementary strand synthesis from the 3'-end of the displaced parental strand through type II replication intermediates. "Pan-handled" structures proposed by DANIELL (1976) as substrates for the initiation of complementary strand synthesis are formed by intramolecular base pairing of the inverted terminal repetitions. At present, they are purely hypothetical and have not been observed and recognized as replicative intermediates. In principle, the duplex positions of pan-handled single strands are identical in structure regardless of whether they are formed from r- or l-strands. They are therefore indistinguishable both from each other and from the molecular ends of intact duplex DNA for the enzymatic machinery responsible for the initiation reaction. This aspect of the model emphasizes the structural and functional equivalence of all initiation events in displacement as well as complementary strand synthesis. It also assumes that the inverted terminal repetitions which form the double-stranded parts of the pan-handled structures rather than additional flanking sequences are the necessary structural prerequisites for viral DNA replication to occur efficiently.

This aspect of the model, which could now be approached experimentally through in vitro replication systems, has not been truly verified. The absence of pan-handled structures in preparations of replicating viral DNA, however, even after in vivo fixation through induction of DNA intrastrand crosslinks, still leaves the possibility of two different initiation mechanisms open (REVET and BENICHOU 1981).

This current view of the mechanism of adenovirus DNA replication is supported by a functional analysis on the location of origins and termini. In these studies both termini (TOLUN and PETTERSSON 1975; SCHILLING et al. 1975; BOUR-

GAUX et al. 1976; HORWITZ 1976; WEINGÄRTNER et al. 1976; SUSSENBACH and KUIJK 1977; ARIGA and SHIMOJO 1978; ARENS and YAMASHITA 1978) and origins (ARIGA and SHIMOJO 1977; SUSSENBACH and KUIJK 1978a; REITER et al. 1980) of viral DNA replication were identified from the asymmetric distribution of newly incorporated radioactivity along the genome and within the two DNA strands of terminal restriction enzyme fragments. They were located approximately within the terminal 260 bp of Ad2 or Ad5 DNA. Recent in vitro replication studies (see below) place the sites of initiation and termination of viral DNA synthesis exactly in the very last nucleotide pairs of the linear double-stranded genome. This is not to say that the origins of DNA replication, i.e., the recognition sites for the proteins which catalyze the particular mechanism of adenovirus DNA replication, are not represented by larger sequence regions located more internally within the viral genome (see below).

Pulse-label studies have also revealed the rate of DNA chain elongation during adenovirus DNA replication. From a replication time of approximately 20 min at 37 °C and a genome length of 36 kb, it can be estimated to be in the order of 1700 bases/min. There are no detectable differences for different serotypes (BODNAR and PEARSON 1980a). Using a density-labeling technique, BODNAR and PEARSON (1980b) have also determined the doubling time for the production of viral DNA during the logarithmic phase of adenovirus replication. With approximately 60 min it is 2.8 times longer than the average time needed for replication of a single adenovirus DNA molecule. It is concluded from those data that initiation rather than elongation is the rate-limiting step in adenovirus replication.

The termini of replicating viral DNA are tightly associated with proteinaceous material. Terminal restriction enzyme fragments are not only retained on BND-cellulose (HORWITZ et al. 1979; STILLMAN and BELLETT 1979) or glass filters (COOMBS et al. 1979; ROBINSON et al. 1979), but also display reduced mobility in agarose gels (VAN WIELINK et al. 1979). A comparison with the same fragments obtained from virions suggested the presence of a protein larger than the 55K protein associated with the termini of virion-derived DNA (FÜT-TERER 1982). It was subsequently identified as the 80K pTP, the relationship between pTP and the 55K TP having been established by partial proteolysis (CHALLBERG et al. 1980) and partial amino acid sequence analysis (SMART and STILLMAN 1982).

5 Mechanism of Initiation of DNA Replication

The absence of terminal redundancies in the linear adenovirus DNA molecule poses a special problem for the initiation of viral DNA replication. Eukaryotic DNA polymerases can only initiate de novo DNA synthesis through a free 3'-hydroxyl group (WEISSBACH 1975). This is usually supplied by a short RNA primer (DRESSLER 1975). The removal of this primer leaves a short single-stranded region at the termini of the replicating DNA molecule which eventually matures by a "fill-in" reaction. This reaction again requires a primer 3'-hydroxyl group. In terminal redundant DNA molecules, this difficulty is simply solved by the formation of circular or concatemeric intermediates (WATSON 1972). For adenovirus DNA which is not terminally redundant several other solutions have been discussed. One of the principal solutions proposed involves the formation of circular double-stranded replication intermediates such as have been discussed for "early" lambda DNA replication, for example (TOMIZAWA and OGAWA 1968). In adenovirus DNA, these structures would not only have to be formed by blunt-end joining of the two termini of the viral DNA; there is also no evidence – as discussed above – for the involvement of such structures in replicating viral DNA.

According to another proposal, adenovirus replication requires the formation of "hairpins" at the 3'-termini of the viral DNA (CAVALIER-SMITH 1974). Yet DNA sequence analysis of the ITRs of adenovirus DNA certainly argues against the presence of foldback sequences and thus against this model (STILLMAN et al. 1977; SUSSENBACH and KUIJK 1978b).

A truly provocative idea has been presented by REKOSH et al. (1977). Their model involves the terminal protein. Through binding to a dCTP-residue and subsequent association of the protein/dCMP complex with the termini of the template DNA, the 3'-OH group of the dCMP residue would provide the necessary primer for DNA replication. This "protein-priming" model, as it has been incorrectly but justifiably termed, is now supported by considerable and convincing experimental evidence. The most important set of data derives from in vitro DNA replication studies. The first in vitro systems for adenovirus DNA replication consisted of isolated nuclei or subnuclear replication complexes from adenovirus-infected cells (SUSSENBACH and VAN DER VLIET 1972; WINNACKER 1975; YAMASHITA et al. 1975; BRISON et al. 1977; KAPLAN et al. 1977). However, these systems were only capable of elongating DNA chains which had already been initiated in vivo prior to the isolation of the nuclei. More recently, however, soluble enzyme systems depending on and replicating exogenously added adenovirus DNA have been described. The first soluble systems consisted of low-salt extracts (100 mM) from nuclei of Ad2- or Ad5-infected HeLa cells (CHALLBERG and KELLY 1979a, b; REITER et al. 1980). More recently it was shown that mixtures of cytoplasmic extracts from infected cells and nuclear extracts from infected or uninfected HeLa cells (KAPLAN et al. 1979; IKEDA et al. 1980; HORWITZ and ARIGA 1981) and a combination of partially purified protein fractions (IKEDA et al. 1981) possess similar activities. Fractionation of the extract from uninfected cells led to the identification of two host factors required for initiation of adenovirus DNA replication (NAGATA et al. 1982, 1983a). Both factors are free of DNA polymerase activity. The 47K nuclear factor I binds between positions 17–48 within the ITR and appears to be required for initiation and early elongation (NAGATA et al. 1983b). Nuclear factor II has a type I DNA topoisomerase activity and permits the synthesis of full-length molecules (NAGATA et al. 1983a).

Three virus-coded polypeptides required for DNA replication were identified in the cytosol from Ad2-infected cells: the 72K adenovirus DBP (KAPLAN et al. 1977, 1979; HORWITZ 1978; LICHY et al. 1981; IKEDA et al. 1981; CHALLBERG et al. 1982; SUSSENBACH and VAN DER VLIET 1982; FRIEFELD et al. 1983a), the

80K (CHALLBERG et al. 1980; STILLMAN et al. 1981; GREEN et al. 1981), and a 140K protein possessing DNA polymerase activity (Pol) (ENOMOTO et al. 1981). Under optimal conditions for viral DNA synthesis, which require the Ad2 DNA/protein complex (1–2 mg/ml) or adenovirus DNA/protein cores (GODING and RUSSELL 1983), Mg^{2+} (5 mM), ATP (3.75 mM), and the dNTPs (50 μM each), the pattern of DNA replication closely resembles the displacement synthesis step observed in vivo. However, it has never been established whether these extracts are able to perform complementary strand synthesis.

The extracts or protein factors have been instrumental in solving the mechanism of initiation of adenovirus DNA replication. Accordingly, the initiation step is defined as the template-dependent formation of a complex between the 80K pTP and dCTP. Three different assay systems have been developed to analyze this reaction. The most direct approach utilizes α-^{32}P-dCTP in an assay which lacks the other three dNTPs. In crude extracts that contain endogenous dNTPs it is necessary to add ddATP or ddTTP to block any chain elongation. After appropriate incubation times the total reaction mixture is boiled in the presence of SDS and subjected to polyacrylamide gel electrophoresis. A ^{32}P-labeled protein band at the 80K position on the autoradiograph indicates a successful initiation reaction (LICHY et al. 1981; PINCUS et al. 1981; STILLMAN et al. 1981; CHALLBERG et al. 1982).

Occasionally this assay has been performed in the presence of α-^{32}P-labeled dCTP, dATP, dTTP, and ddGTP. The DNA chain is then elongated up to position 26, the location of the first dG residue in the adenovirus sequence. The resulting complex between the 80K protein (pTP) and the short oligonucleotide (a 26-mer) migrates upon polyacrylamide gel electrophoresis as an 88K band (LICHY et al. 1981). A third assay uses restriction enzyme fragments of the adenovirus DNA/protein complex as a substrate. Following incubation under optimal conditions for DNA synthesis in the presence of α-^{32}P-labeled dNTPs these fragments become labeled in their DNA positions and are subsequently separated upon agarose gel electrophoresis. With extracts active for the initiation reaction, only terminal restriction enzyme fragments and not internal fragments will become labeled (HORWITZ and ARIGA 1981; STILLMAN et al. 1982b).

Crude in vitro systems often contain DNA-nicking activities, which subsequently permit a repair-type synthesis by using endogenous DNA polymerases. Under these circumstances internal fragments may also become labeled. This undesired reaction can be suppressed by low concentrations of aphidicolin, indicating the participitation of the cellular DNA Polymerase α in these repair processes (PINCUS et al. 1981; LICHY et al. 1981).

In the presence of terminal restriction fragments single-stranded DNA derived from these fragments will also become labeled in the initiation assay. These single strands originate from multiple initiation events occurring at the termini of the double-stranded fragments (HORWITZ and ARIGA 1981; STILLMAN et al. 1982b; MEINSCHAD 1983). In contrast to the intact viral genome, these fragments contain only one proper terminus, and therefore single strands of only one polarity (the r-strand from left-ended fragments and vice versa) are constantly synthesized and released into the assay mixture. Due to their different mobility with regard to their double-stranded counterparts, they can easily be recognized as faster-moving bands on the autoradiographs. Since the first initiation event only releases an unlabeled single strand derived from the unlabeled parental fragment the appearance of labeled single strands can only be due to at least two or multiple initiation events. In contrast to the assay which labels the 80K pTP/dCMP complex, this latter assay does not identify the initiation event directly, but it is specific for a proper and active initiation system and at the same time permits an analysis of the elongation step. In conjugation with an internal control for proper initiation this elongation assay can thus be used to study the enzymatic requirements of displacement DNA synthesis.

In all three assay systems the optimal substrate is a viral DNA/55K protein complex isolated from virions by denaturation in 4 *M* guanidine hydrochloride. It is purified through sedimentation in sucrose gradients containing 4 *M* guanidine hydrochloride (SHARP et al. 1976) and contains the terminally linked 55K terminal protein (TP). If this is only partially removed by proteolysis, e.g., with proteinase K, the template is inactive (LICHY et al. 1981; TAMANOI and STILLMAN 1982). However, a complete removal of the protein through alkali treatment, e.g., piperidine (STILLMAN et al. 1981; TAMANOI and STILLMAN 1982), fully restores the activity. Thus the presence of the parental TP on the 5'-end of the displaced strand is not required for the initiation reaction. This point is particularly well demonstrated by the use of linearized plasmid DNA fragments. TAMANOI and STILLMAN (1982) have cloned the left-hand terminus of Ad2 DNA by digestion of the intact Ad2 DNA/protein complex with the restriction enzyme *Bgl*II, followed by removal of the terminal protein through piperidine treatment and by subsequent ligation of the blunt-ended fragment with an *Eco*RI-linker (GGAATTCC). After cloning in pBR322 the recombinant plasmid (pLA1) could be linearized, creating one terminus identical with Ad2 DNA apart from two G/C base pairs at the 5'-overhanging *Eco*RI recognition sequence (Fig. 5). These linear DNA molecules show activity in forming the 80K pTP/dCMP initiation complex even after heat denaturation. Similar results have been obtained using a cloned terminal *Xba*I DNA fragment with no extra sequences apart from the overhanging *Eco*RI ends (VAN BERGEN et al. 1983). These results convincingly prove that it is not the terminal protein on the displaced parental strand but rather a particular sequence requirement that is needed for the proper initiation reaction. The possibility of using cloned terminal fragments as templates in the initiation assay certainly provides the opportunity to study the structural and sequence requirements of this reaction. The available data (Fig. 5) show that the presence of two additional C/G base pairs at the terminus in addition to the *Eco*RI overhanging ends is tolerated (pLA1/*Eco*RI), while three or more G/C base pairs block complex formation (pWG/*Eco*RI; *8/*Eco*RI) (TAMANOI and STILLMAN 1982; LALLY et al. 1984). On the other hand, removal of up to eight base pairs from the terminus does not eliminate the activity. This indicates that the highly conserved sequence between positions 9 and 17 might play a critical role in the initiation process (VAN BERGEN et al. 1983). This was demonstrated by the chemical synthesis of terminal sequences 17 bp, 19 bp, 21 bp and 38 bp long. After cloning in pBR327 and subsequent linearization, the 19 bp, 21 bp and 38 bp long termini were able to catalyze

Template		Replication

```
        ↓   ↓
    5'AATTCCCATCATCAA- - -        pLAI / EcoRI          + +
     3'GGGTAGTAGTT- - -
```

```
        ↓   ↓
    5'AATTCATCATCAA- - -          XD7 / Eco RI          + +
     3'GTAGTAGTT- - -
```

```
    5'AATTCGGGCATCATCAA- - -      pWG / EcoRI           - -
     3'CCCGTAGTAGTT- - -
```

```
    5'GG- - - - -GGCATCATCAA  - -    *8 / Pst I           - -
   3'ACGTCC- - - - -CCGTAGTAGTT- - -
```

Fig. 5. Sequence requirements for the initiation of adenovirus DNA replication in vitro. Terminal restriction fragments of Ad2 were cloned by various methods. After linearization with appropriate restriction enzymes the plasmids were used as templates in the in vitro replication assay described in the text. *Arrows* indicate positions where new strands can be initiated. Data are taken from TAMANOI and STILLMAN (1982) (pLAI, *8), VAN BERGEN et al. (1983) (XD7), and LALLY et al. (1984) (pWG)

the in vitro initiation reaction while the 17 bp terminus was totally inactive (LALLY et al. 1984). In addition, site-directed mutagenesis within the conserved region between positions 9 and 17 destroyed all template activity (TAMANOI and STILLMAN 1983; DÖRPER and WINNACKER, to be published). These results indicate that the terminal 18 to 19 base pairs constitute a functional origin for the initiation of adenovirus DNA replication in vitro. A different approach has been taken by LALLY et al. (1984) who have used heterologous in vitro systems to define the sequence requirements of the initiation reaction. The first 17 terminal base pairs of the mouse AdFl genome and the human Ad2 and Ad5 genomes are identical (TEMPLE et al. 1981). The AdFl DNA/protein complex can be replicated efficiently in an in vitro system derived from nuclear extracts of AdFl-infected 3T3 (mouse) cells. However, it is entirely inactive in an in vitro system derived from Ad2-infected HeLa cells; and conversely the Ad2 DNA/protein complex is inactive in the in vitro system from mouse cells. The first 17 bp that both virus genomes have in common are thus not sufficient to be recognized by a heterologous replication system. The complexes between the respective terminal proteins and dCMP apparently require other or additional sequence determinants to be active as templates in adenovirus DNA synthesis.

It remains to be seen whether – as is observed for the adenovirus DNA/TP complexes – the minimalized plasmid replicons can also be replicated in vivo and whether there are different sequence requirements in this situation. It is interesting in this context that the ITRs of primate adenoviruses share homologies not only in the AT-rich terminal region but also in the more internal GC-rich position. This does not only refer to the sequence GGGPyGGPuG, which is also present in the origins of human BK virus (SEIF et al. 1979), SV40 (SUBRAMANIAN et al. 1977), and polyoma virus (SEIF et al. 1979), but also to the highly conserved hexanucleotide sequence TGACGT located close to and around the end of the ITRs (TOLUN et al. 1979; STILLMAN et al. 1982a). Whether these sequences are of any particular significance for the initiation reaction will certainly be evaluated in the near future.

No experiments have been performed specifically on the mechanism of initiation of complementary strand synthesis, athough the necessary experimental means are available.

6 Adenovirus DNA Replication Proteins

So far, five proteins have been identified as being required for adenovirus DNA replication (STILLMAN 1983). Three are virus-coded and two are provided by the host genome. Among the three virus-coded proteins are the 72K DBP, the 80K pTP, and a 140K protein possessing DNA polymerase activity. The latter two proteins can be found and isolated as a pTP-pol complex (ENOMOTO et al. 1981), which binds specifically to the termini of adenovirus DNA between positions 9 and 22 (RIJNDERS et al. 1983b). This region barely overlaps with a stretch of 31 base pairs between position 17 and 48 which is protected from DNAse I digestion through the interaction with nuclear factor I (NAGATA et al. 1983b).

The pTP-pol complex is separated into its subunits by sedimentation through a glycerol gradient in the presence of urea (LICHY et al. 1982; STILLMAN et al. 1982b). The largest subunit, the 140K protein, possesses an aphidicolin-resistant DNA polymerase activity (ENOMOTO et al. 1981). This activity was shown to complement replication (initiation)-defective extracts prepared from Ad5ts149-infected HeLa cells (STILLMAN et al. 1982b). The mutant Ad5ts149 (GINSBERG et al. 1974), together with Ad5ts36 (WILKIE et al. 1973), belongs to the complementation group N of temperature-sensitive mutants which are unable to replicate viral DNA at nonpermissive temperature (VAN DER VLIET and SUSSENBACH 1975; CARTER and GINSBERG 1976) and are unable to transform rat embryo cells under similar conditions (WILLIAMS et al. 1974, 1979). Both mutants map between coordinates 18 and 22 from the left end of the viral genome (GALOS et al. 1979) within an open reading frame extending from coordinate 22.9 leftward to coordinate 14.2 (ALESTRÖM et al. 1982a; GINGERAS et al. 1982). This area does not overlap with the open translational reading frame encoding pTP (Fig. 3). Group N mutants thus constitute a new complementation group. The complementation of replication-defective extracts from Ad5ts149-infected cells in the pTP/dCMP assay with a 140K DNA polymerase activity from wild-type virus infected cells definitively proves that adenovirus encodes a novel DNA polymerase activity that is required for priming of DNA synthesis at the origin of DNA replication (STILLMAN et al. 1982b). At present it is not clear whether the 140K protein is also functioning in chain elongation. Previous studies have suggested (BOLDEN et al. 1975; DE JONG et al. 1977) that adenovirus does not induce DNA polymerase activity in infected cells and thus relies entirely on host polymerases. All three cellular polymerases have been implicated in these functions; polymerase α on the basis of a marginal sensitivity of adenovirus DNA replication to aphidicolin (IKEGAMI et al. 1978; KROKAN et al. 1979; LONGIARU et al. 1979; WIST and PRYDZ 1979; VAN DER WERF et al. 1980), polymerase β (IKEDA et al. 1980, 1981) and polymerase γ because dideoxy NTPs, at ratios to dNTPs that do not affect SV40 DNA synthesis, strongly inhibit adenovirus DNA replication (KROKAN et al. 1979; SHAW et al. 1980; KWANT and VAN DER VLIET 1980; VAN DER WERF et al. 1980; HABARA et al. 1980). The discovery of the 140K gene N product possessing properties different from all three host polymerases with respect to template preference and sensitivity to inhibitors (ENOMOTO et al. 1981; LICHY et al. 1982) solves the apparent confusion. However, the sensitivity of adenovirus DNA replication towards aphidicolin in vivo still leaves open the participition of a cellular DNA polymerase activity in chain elongation.

As described above, in vitro studies suggest that the block in DNA synthesis for Ad5ts149 is clearly at the stage of initiation (LICHY et al. 1982; STILLMAN et al. 1982b; OSTROVE et al. 1983; VAN BERGAN and VAN DER VLIET 1983). However, in vivo DNA replication declines only slowly over the course of 4–8 h following a shift to the nonpermissive temperature (CARTER and GINSBERG 1976), although it takes only 30 min to complete one round of viral DNA synthesis. These conflicting results can be reconciled if one assumes that the defect in Ad5ts149 represents the lack of association of the 140K polypeptide with pTP at the nonpermissive temperature, thereby preventing the formation of the initiation complex. A complex formed at the permissive temperature, however, may be refractory to a shift to the high temperature, allowing initiation to occur until the supply of the 140K polymerase is exhausted (STILLMAN et al. 1982b). FRIEFELD et al. (1983b) have shown that the mutant pTP/Pol complex dissociates more rapidly in the presence of urea than the wild-type complex. This could be due either to an altered polymerase activity that is less strongly associated with pTP than the wild-type enzyme or to a reduction in the amount of polymerase molecules. In any case, these observations can certainly explain the in vivo and in vitro replication data and are consistent with a role of the gene N product in initiation of viral DNA replication.

Apart from the (80K) pTP/Pol complex, the third virus-coded protein is the adenovirus DNA-binding protein (DBP). It is coded for by region E2A between coordinates 61.6 and 66.5 on the viral l-strand and has an apparent molecular weight of 72K daltons (Fig. 6). However, a molecular weight of 59 048 daltons of the 529 amino acid-long polypeptide could be derived from nucleotide sequence analysis (KRUIJER et al. 1981, 1982). The in vitro system described and the existence of two ts mutants (Ad5ts125 and Ad5ts107) have been instrumental in defining several independent functions of this protein.

First the 72K DBP binds strongly but not covalently to single-stranded DNA (VAN DER VLIET and LEVINE 1973; VAN DER VLIET et al. 1977; FOWLKES et al. 1979; SCHECHTER et al. 1980; NASS and FRENKEL 1980). The ts alterations in H5ts125 or H5ts107 diminish the DNA-binding activity of the protein in vitro (ENSINGER and GINSBERG 1972; VAN DER VLIET et al. 1975) and inhibit DNA replication in vivo (VAN DER VLIET and SUSSENBACH 1975; VAN DER VLIET et al. 1977). With cytosol extracts from mutant-infected cells it could be shown

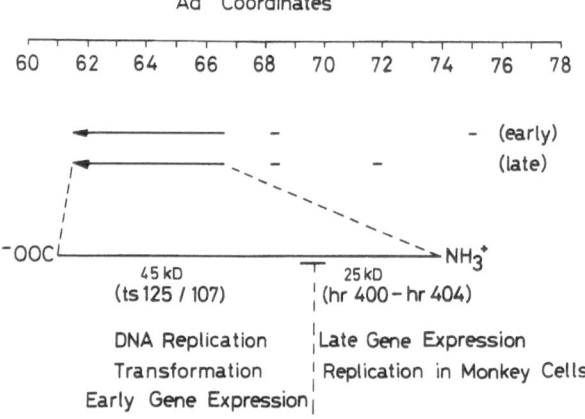

Ad Coordinates

60   62   64   66   68   70   72   74   76   78

45 kD
(ts 125 / 107)

25 kD
(hr 400 – hr 404)

DNA Replication   Late Gene Expression
Transformation   Replication in Monkey Cells
Early Gene Expression

Fig. 6. Functional domains of the DNA-binding 72K protein. The protein is encoded between positions 61.5 and 66.5 on the viral l-strand (KRUIJER et al. 1982). Analysis of the phenotypes of various *ts* and *hr* mutants as indicated localizes different functions of the 72K protein to either of two chymotryptic fragments, a 45K C-terminal and a 25K N-terminal fragment. The exact positions of the *ts* mutants on the genome have been determined by DNA sequencing (KRUIJER et al. 1983)

that DBP is not essential for the initiation reaction, e.g., for the formation of the 80K pTP/Pol complex (CHALLBERG et al. 1982; SUSSENBACH and VAN DER VLIET 1982). Rather it is necessary in elongation of Ad DNA synthesis (KAPLAN et al. 1977, 1979; HORWITZ et al. 1979). Although the synthesis of an early elongation product consisting of the pTP and a short oligonucleotide 26 nucleotides long can proceed in the absence of DBP, it is stimulated significantly by the presence of a functional DBP (LICHY et al. 1981; IKEDA et al. 1981; FRIEFELD et al. 1983a; VAN BERGEN and VAN DER VLIET 1983). In this case a host DBP may substitute for the adenovirus DBP. More extended elongation reactions are strictly dependent on adenovirus DBP, however.

A second function relates to the turn-off of viral "early" transcription upon the onset of viral DNA replication. Cells infected at nonpermissive temperature (39 °C) with Ad5*ts*125 accumulate early viral RNA due to an increased stability of viral mRNA (NEVINS and JENSEN-WINKLER 1980; BABICH and NEVINS 1981). The inhibitory effect of the DBP on adenovirus "early" transcription has been specifically related to region EIV and could also be demonstrated in vitro using a whole cell extract in a "run-off" assay (Handa et al. 1983).

The third function of the adenovirus DBP involves transformation. At the nonpermissive temperature the *ts* mutants transform rat cells in culture with a higher efficiency than wild-type virus (WILLIAMS et al. 1974). The fourth function of Ad DBP is defined by a class of mutants which enhance the ability of human Ad2 to grow in monkey cells. In monkey cells infected with wild-type Ad2 a block of late gene expression, in particular the fiber protein message, has been related to a defect in the processing of fiber mRNA (KLESSIG and CHOW 1980). This defect can be overcome by mutants in the DBP.

Biochemical and genetic studies have shown that these mutants reside in different domains of this protein. The mutations which affect DNA replication and binding to single-stranded DNA map in the C-terminal half of the molecule. Both Ad5ts125 and Ad5ts107 show the same alteration in amino acid 413 from proline to serine (KRUIJER et al. 1983). In contrast, the alterations in the host-range mutants that allow the virus to grow on monkey cells are all located in the N-terminal half of the DBP. Since these mutants have no effect on DNA replication or early gene expression, these data suggest the existence of a separate functional domain essential for DNA replication. Chymotryptic digestion produces a 25K heavily phosphorylated N-terminal and a 45K C-terminal fragment with very few phosphorylation sites (KLEIN et al. 1979; LINNÉ and PHILIPSON 1980; Fig. 6). This C-terminal fragment (ARIGA et al. 1980) and an even smaller 34K fragment derived from it retain full activity in in vitro elongation of Ad DNA. Since the 34K C-terminal subfragment is completely free of phosphate residues, phosphorylation does not appear to be necessary for DNA replication (FRIEFELD et al. 1983a) The exact location of the other functions will undoubtedly be available in the near future.

The multifunctionality of the adenovirus DBP is reminiscent of the situation in SV40 large T antigen or even the T4 gene 32 product (MOSIG et al. 1977). It is interesting to note in this context that SV40 T antigen, or even a small subfragment from its C terminus, can complement the deficiency of adenovirus to grow in monkey cells (LEWIS et al. 1969; FEY et al. 1979), just as mutations in the adenovirus DBP itself extend the host range of Ad2 to these cells (KLESSIG and HASSEL 1978; KLESSIG and GRODZICKER 1979).

The range of interaction of adenoviruses with different host cells appears exceedingly complex. Human Ad2 and Ad5 replicate efficiently not only in human HeLa and KB cells, but also in hamster cells (SHIMOJO and YAMASHITA 1968), as is also documented by the activity of extracts from adenovirus-infected CHO cells in in vitro replication experiments (LONGIARU and HORWITZ 1981). In contrast, monkey, rat and mouse cells are only semipermissive, yielding titers of infectious progeny which are at least 2–3 orders of magnitude lower than those produced in human KB cells. Even within one species, e.g., the mouse, considerable differences in permissiveness are observed depending on the state of cell differentiation. Thus the F9 line of mouse teratocarcinoma cells, which does not differentiate under normal culture conditions, produces 100-fold lower yields than an embryonal carcinoma line that spontaneously undergoes differentiation in vitro (CHENG and PRASZKIER 1982). The nature of the block in adenovirus synthesis under abortive conditions has best been studied in the monkey system. In AGMK cells the observed block to replication of Ad2 is a late event. The synthesis of the major early mRNA species, of early viral proteins, and of viral DNA appears to be unaffected. However, the synthesis of most late viral proteins is reduced. In the case of fiber protein the drastic reduction of fiber protein synthesis (100- to 1000-fold) is accompanied only by a moderate reduction of the respective mRNA levels (KLESSIG and CHOW 1980).

This mRNA is just as translatable in vitro as the fiber message from productively infected cells. Although the exact nature of the block is not known, it has been suggested (ANDERSON and KLESSIG 1983) that it is associated with a defect in the stability of mRNA/ribosome complexes in abortively infected cells.

7 Conclusions

Since the most recent reviews were completed (WINNACKER 1978; CHALLBERG and KELLY 1982), considerable advances have been achieved in our understanding of the mechanism of adenovirus DNA replication. The establishment of in vitro systems for de novo initiation, the cloning of the adenovirus termini, and the observation that cloned termini can be active in an in vitro replication system have been landmarks in this development. It has opened up new areas of research, e.g., the incipient attempts to minimize the adenovirus replicon and the recent approaches to the use of the adenovirus replicon in certain eukaryotic vectors. Major advances are to be expected in the enzymology of adenovirus DNA replication. The discovery of nuclear, host-derived factors in the initiation reaction has raised anew the question of the involvement of host gene products in the regulation of virus/cell interaction. Together with the studies on adenovirus interactions with abortively infected cells and with the employment of new techniques in molecular cloning it is possible that this exciting field of work might, as has happened in the area of eukaryotic DNA replication, again see the adenovirus DNA molecule as a model system. Whether this will keep us all busy for another 30 years remains to be seen.

Acknowledgments. We thank C. LALLY for critically reading the manuscript, and G. KAUSEL and S. PREGLER for their help in preparing the manuscript. Experiments carried out in the laboratory of E.-L. WINNACKER were supported by the Deutsche Forschungsgemeinschaft.

References

Akusjärvi G, Persson H (1981) Gene and mRNA for precursor polypeptide VI from adenovirus type 2. J Virol 38:469–482

Aleström P, Akusjärvi G, Pettersson M, Pettersson U (1982a) DNA sequence analysis of the region encoding the terminal protein and the hypothetical N-gene product of adenovirus type 2. J Biol Chem 257:13492–13498

Aleström P, Stenlund A, Li P, Pettersson U (1982b) A common sequence in the inverted terminal repetitions of human and avian adenoviruses. Gene 18:193–197

Anderson KP, Klessig DF (1983) Posttranscriptional block to synthesis of a human adenovirus capsid protein in abortively infected monkey cells. J Mol Appl Gen 2:31–43

Antoine G, Aleström P, Schilling R, Pettersson U, Winnacker E-L (1982) Defective expression of mouse adenovirus Fl in human cells. EMBO J 1:453–459

Arens M, Yamashita T (1978) In vitro termination of adenovirus DNA synthesis by a soluble replication complex. J Virol 25:698–702

Ariga H, Shimojo H (1978) Initiation and termination of adenovirus 12 DNA replication. II. Analysis of pulse-labeled oligonucleotides derived from 5'-termini in the DNA strand. Virology 85:58–108

Ariga H, Klein H, Levine AJ, Horwitz MS (1980) A cleavage product of the adenovirus DNA binding protein is active in DNA replication in vitro. Virology 101:307–310

Arrand JR, Roberts RJ (1979) The nucleotide sequences at the termini of Ad2 DNA. J Mol Biol 128:577–594

Babich A, Nevins JR (1981) The stability of early adenoviral mRNA is controlled by the viral 72 kD DNA-binding protein. Cell 26:371–379

Bodnar JW, Pearson GD (1980a) Kinetics of adenovirus DNA replication. I. Rate of adenovirus DNA replication. Virology 100:208–211

Bodnar JW, Pearson GD (1980b) Kinetics of adenovirus DNA replication. II. Initiation of adenovirus DNA replication. Virology 105:357–370

Bolden A, Aucker J, Weissbach A (1975) Synthesis of herpes simplex virus, vaccinia virus and adenovirus DNA in isolated HeLa cell nuclei. J Virol 16:1584–1592

Bourgaux P, Delbecchi L, Bourgaux-Ramoisy D (1976) Initiation of adenovirus type 2 DNA replication. J Virol 72:89–98

Brinckmann U, Darai G, Flügel RM (1984) Tupaia (Tree shrew) adenovirus DNA: Sequence of the left-hand fragment corresponding to the transforming early region of human adenoviruses. Gene 24:131–135

Brison O, Kedinger C, Wilhelm J (1977) Enzymatic properties of viral replication complexes isolated from adenovirus type 2 infected HeLa cell nuclei. J Virol 24:423–443

Brown DT, Westphal M, Burlingham BT, Winterhoff U, Doerfler W (1975) Structure and composition of the adenovirus type 2 core. J Virol 16:366–387

Brown M, Weber J (1980) Virion core-like organization of intranuclear adenovirus chromatin late in infection. Virology 107:306–310

Carter TH, Ginsberg P (1976) Viral transcription in KB cells infected by temperature sensitive early mutants of adenovirus type 5. J Virol 18:156–166

Cavalier-Smith T (1974) Palindromic base sequences and replication of eukaryotic chromosome ends. Nature 250:467–470

Challberg MD, Kelly TJ Jr (1979a) Adenovirus DNA replication in vitro. Proc Natl Acad Sci USA 76:655–659

Challberg MD, Kelly TJ Jr (1979b) Adenovirus DNA replication in vitro. Origin and direction of daughter strand synthesis. J Mol Biol 135:999–1011

Challberg MD, Kelly TJ Jr (1981) Processing of the adenovirus terminal protein. J Virol 38:272–277

Challberg MD, Kelly TJ Jr (1982) Eucaryotic DNA replication: viral and plasmid model systems. Annu Rev Biochem 51:901–934

Challberg MD, Desiderio SV, Kelly TJ Jr (1980) Adenovirus DNA replication in vitro: characterization of a protein covalently linked to nascent DNA strands. Proc Natl Acad Sci USA 77:5105–5109

Challberg MD, Ostrove JM, Kelly TJ Jr (1982) Initiation of adenovirus DNA replication: Detection of covalent complexes between nucleotide and the 80 kD terminal protein. J Virol 41:265–270

Cheng C, Praszkier J (1982) Regulation of type 5 adenovirus replication in murine teratocarcinoma cell lines. Virology 123:45–59

Chinnadurai G, Chinnadurai S, Green M (1978) Enhanced infectivity of adenovirus type 2 DNA and a DNA-protein complex. J Virol 26:195–199

Coombs DH, Robinson AJ, Bodnar JW, Jones CJ, Pearson GD (1979) Detection of DNA-protein complexes: the adenovirus DNA-terminal protein and HeLa DNA-protein complexes. Cold Spring Harbor Symp Quant Biol 43:741–753

Corden J, Engelking HM, Pearson GD (1976) The chromatin-like organization of the adenovirus chromosome. Proc Natl Acad Sci USA 73:401–404

Daniell E (1976) Genome structure of incomplete particles of adenovirus. J Virol 19:685–708

Daniell E, Groff DE, Fedor MJ (1981) Adenovirus chromatin structure at different stages of infection. Mol Cell Biol 1:1094–1105

De Jong A, van der Vliet PC, Jansz HS (1977) DNA polymerases in adenovirus type 5 infected and uninfected KB cells. Induction of an α-type DNA polymerase in adenovirus type 5 infected and fast growing cells. Biochim Biophys Acta 476:156–165

De Jong PJ, Kwant MM, van Driel W, Jansz HS, van der Vliet PC (1983) The ATP requirements of adenovirus type 5 DNA replication and cellular DNA replication. Virology 124:45–58

Desiderio SV, Kelly TJ Jr (1981) Structure of the linkage between adenovirus DNA and 55000 molecular weight terminal protein. J Mol Biol 145:319–337

Dijkema R, Dekker BMM (1979) The inverted terminal repetition of the DNA of weakly oncogenic adenovirus type 7. Gene 8:7–15

Dressler D (1975) The recent excitement in the DNA growing point problem. Ann Rev Microbiol 29:525–579

Ellens DJ, Sussenbach JS, Jansz HS (1974) Studies on the mechanism of replication of adenovirus DNA. III. Electron microscopy of replicating DNA. Virology 61:427–442

Enomoto T, Lichy JH, Ikeda J, Hurwitz J (1981) Adenovirus DNA replication in vitro: purification of the terminal protein in a functional form. Proc Natl Acad Sci 78:6779–6783

Ensinger JM, Ginsberg HS (1972) Selection and preliminary characterization of temperature-sensitive mutants of type 5 adenovirus. J Virol 10:328–339

Estes MK (1978) Characterization of DNA-protein complexes from simian adenovirus SA7. J Virol 25:917–922

Fanning E, Doerfler W (1977) Intracellular forms of adenovirus DNA. Virology 81:433–448

Fey G, Weqis JB, Grodzicker T, Bothwell A (1979) Characterization of a fused protein specified by the adenovirus type 2 simian virus 40 hybrid Ad2 + ND1dp2. J Virol 30:201–217

Flint SJ, Berget SM, Sharp PA (1976) Characterization of single-stranded viral DNA sequences present during replication of adenovirus type 2 and 5. Cell 9:559–571

Flint SJ, Broker TR (1980) Lytic infection by adenoviruses. In: Tooze J (ed) Molecular biology of tumor viruses, part 2. Cold Spring Harbor, New York, p 443

Fowlkes DM, Lord ST, Linné T, Pettersson U, Philipson L (1979) Interaction between the adenovirus DNA-binding protein and double-stranded DNA. J Mol Biol 132:163–180

Friefeld BR, Krevolin MD, Horwitz MS (1983a) Effects of the adenovirus H5ts125 and H5ts107 DNA-binding proteins on DNA replication in vitro. Virology 124:380–389

Friefeld BR, Lichy JH, Hurwitz J, Horwitz MS (1983b) Evidence for an altered adenovirus DNA polymerase in cells infected with the mutant H5ts149. Proc Natl Acad Sci USA 80:1589–1593

Fütterer J (1982) DNA-Protein-Komplexe in Adenovirus-infizierten Zellen. PhD Thesis, Institut für Biochemie, Munich

Galibert F, Herisse J, Courtois G (1979) Nucleotide-sequence of the EcoRI-F fragment of adenovirus 2 genome. Gene 6:1–22

Galos RS, Williams J, Binger M-H, Flint SJ (1979) Location of additional early gene sequences in the adenoviral chromosome. Cell 17:945–956

Garon CF, Parr RP, Padmanabhan R, Robinson I, Garrison JW, Rose J (1982) Structural characterization of the adenovirus 18 inverted terminal repetition. Virology 121:230–239

Gingeras TR, Sciaky D, Gelinas RE, Bing-Dong J, Yen CE, Kelly MM, Bullock PA, Parsons BL, O'Neill KE, Roberts RJ (1982) Nucleotide sequences from the adenovirus 2 genome. J Biol Chem 257:13475–13491

Ginsberg HS, Ensinger MJ, Kauffmann RS, Mayer AJ, Lundholm U (1974) Cell transformation: A study of regulation with types 5 and 12 adenovirus temperature-sensitive mutants. Cold Spring Harbor Symp Quant Biol 39:419–426

Goding CR, Russell WC (1983) Adenovirus cores can function as templates in in vitro DNA replication. EMBO J 2:339–344

Green M, Mackey JK, Wold WSM, Rigden P (1979) Thirty-one human adenovirus serotypes (Ad1-31) form five groups (A–E) based upon genome homologies. Virology 93:481–492

Green M, Symington J, Brackmann KH, Cartas MA, Thornton H, Young L (1981) Immunological and chemical identification of intracellular forms of adenovirus type 2 terminal protein. J Virol 40:541–550

Habara A, Kano K, Nagano H, Mano Y, Ikegami S, Yamashita T (1980) Inhibition of DNA synthesis in the adenovirus DNA replication complex by aphidicolin and 2′,3′-dideoxythymidine triphosphate. Biochem Biophys Res Commun 92:8–12

Handa H, Kingston RE, Sharp PA (1983) Inhibition of adenovirus early region IV transcription in vitro by a purified viral DNA binding protein. Nature 302:545–547

Herisse J, Galibert F (1981) Nucleotide sequence of the EcoRI-E fragment of adenovirus 2 genome. Nucleic Acids Res 9:1229–1240

Herisse J, Courtois G, Galibert F (1980) Nucleotide sequence of the EcoRI-D fragment of adenovirus 2 genome. Nucleic Acids Res 8:2173–2192

Horwitz MS (1976) Bidirectional replication of adenovirus type 2 DNA. J Virol 18:307–315

Horwitz MS (1978) Temperature-sensitive replication of H5ts125 adenovirus DNA in vitro. Proc Natl Acad Sci USA 75:4291–4295

Horwitz MS, Ariga H (1981) Multiple rounds of adenovirus DNA replication in vitro. Proc Natl Acad Sci USA 78:1476–1477

Horwitz MS, Kaplan LM, Abboud MA, Maritato J, Chow LT, Broker TR (1979) Adenovirus DNA replication in soluble extracts of infected cell nuclei. Cold Spring Harbor Symp Quant Biol 43:769–780

Ikeda JE, Longiaru M, Horwitz MS, Hurwitz J (1980) Elongation of primed DNA templates by eukaryotic DNA polymerases. Proc Natl Acad Sci USA 77:5827–5831

Ikeda JE, Enomoto T, Hurwitz J (1981) Replication of adenovirus DNA-protein complex with purified proteins. Proc Natl Acad Sci USA 78:884–888

Ikegami S, Taguchi T, Ohashi M, Oguro M, Nagano H, Mano Y (1978) Aphidicolin prevents mitotic cell divison by interfering with the activity of DNA polymerase. Nature 275:458–460

Kaplan LM, Kleinmann RE, Horwitz MS (1977) Replication of adenovirus type 2 DNA in vitro. Proc Natl Acad Sci USA 74:4425–4429

Kaplan LM, Ariga H, Hurwitz J, Horwitz MS (1979) Complementation of the temperature-sensitive defect in H5ts125 adenovirus DNA replication in vitro. Proc Natl Acad Sci USA 76:5534–5538

Kelly TJ Jr, Lechner RL (1979) The structure of replicating adenovirus DNA molecules: characterization of DNA-protein complexes from infected cells. Cold Spring Harbor Symp Quant Biol 43:721–728

Klein H, Maltman W, Levine AJ (1979) Structure function relationships of the adenovirus DNA-binding protein. J Biol Chem 254:11051–11060

Klessig DF, Chow LT (1980) Incomplete splicing and deficient accumulation of the fibre messenger RNA in monkey cells infected by human adenovirus type 2. J Mol Biol 139:221–242

Klessig DF, Grodzicker T (1979) Mutations that allow human Ad2 and Ad5 to express late genes in monkey cells map in the viral gene encoding the 72 K DNA-binding protein. Cell 17:957–966

Klessig DF, Hassell JA (1978) Characterization of a variant of human adenovirus type 2 which multiplies efficiently in simian cells. J Virol 28:945–956

Krokan H, Schaffer P, De Pamphilis ML (1979) Involvement of eucaryotic deoxyribonucleic acid polymerases α and γ in the replication of cellular and viral deoxyribonucleic acid. Biochemistry 18:4431–4443

Kruijer W, van Schaik FMA, Sussenbach JS (1981) Structure and organization of the gene coding for the DNA-binding protein of adenovirus type 5. Nucleic Acids Res 9:4439–4457

Kruijer W, van Schaik FMA, Sussenbach JS (1982) Nucleotide sequence of the gene encoding adenovirus type 2 DNA binding protein. Nucleic Acid Res 10:4493–4500

Kruijer W, Nicolas JC, van Schaik FMA, Sussenbach JS (1983) Structure and Function of DNA binding proteins from revertants of adenovirus type 5 mutants with a temperature-sensitive DNA replication. Virology 124:425–433

Kwant MMK, van der Vliet PC (1980) Differential effect of aphidicolin on adenovirus DNA synthesis and cellular DNA synthesis. Nucleic Acids Res 8:3993–4007

Lally C, Dörper T, Gröger W, Antoine G, Winnacker EL (1984) A site analysis of the adenovirus replicon. EMBO Journal 3:333–337

Lavelle G, Patch C, Khoury G, Rose J (1975) Isolation and partial characterization of single-stranded adenoviral DNA produced during synthesis of adenovirus type 2 DNA molecules. J Virol 16:775–782

Lechner RL, Kelly TJ Jr (1977) The structure of replicating adenovirus 2 DNA molecules. Cell 12:1007–1020

Lewis AM Jr, Levin MJ, Wiese WH, Crumpacker CS, Henry PH (1969) A nondefective (competent) adenovirus-SV40 hybrid isolated from the Ad2-SV40 hybrid population. Proc Natl Acad Sci USA 63:1128–1135

Lichy JH, Horwitz MS, Hurwitz J (1981) Formation of a covalent complex between the 80000 dalton adenovirus terminal protein and 5'-dCMP in vitro. Proc Natl Acad Sci USA 78:2678–2682

Lichy JH, Nagata K, Friefeld BR, Enomoto T, Field J, Guggenheimer RA, Ikeda J, Horwitz MS, Hurwitz J (1982) Isolation of proteins involved in the replication of adenovirus DNA in vitro. Cold Spring Harbor Symp Quant Biol 47:731–740

Linné T, Philipson L (1980) Further characterization of the phosphate moiety of the adenovirus type 2 DNA-binding protein. Eur J Biochem 103:259–270

Lischue MA, Sung MT (1977) A histone-like protein from adenovirus chromatin. Nature 267:552–555

Longiaru M, Horwitz MS (1981) Chinese hamster ovary cells replicate adenovirus deoxyribonucleic acid. Mol Cell Biol 1:208–215

Longiaru M, Ikeda J, Jarkowsky Z, Horwitz SB, Horwitz MS (1979) The effect of aphidicolin on adenovirus DNA synthesis. Nucleic Acids Res 6:3369–3386

Meinschad C (1983) Replikation und Rekombination bei Adenoviren. PhD Thesis, Institut für Bio-
    chemie, Munich
Mosig G, Breschkin A, Dannenberg R, Bock S (1977) Multiple interactions of a DNA-binding
    protein gene 32 protein of phage T4 during DNA replication and recombination. In: Molineux
    I, Koliyuma M (eds) DNA synthesis, present and future. Plenum, New York, pp 367–380
Nagata K, Guggenheimer RA, Enomoto T, Lichy JH, Hurwitz J (1982) Adenovirus DNA replication
    in vitro: identification of a host factor that stimulates synthesis of the preterminal protein-dCMP
    complex. Proc Natl Acad Sci USA 79:6438–6442
Nagata K, Guggenheimer RA, Hurwitz J (1983a) Adenovirus DNA replication in vitro: synthesis
    of full-length DNA with purified proteins. Proc Natl Acad Sci USA 80:4266–4270
Nagata K, Guggenheimer RA, Hurwitz J (1983b) Specific binding of a cellular DNA replication
    protein to the origin of replicating DNA. Proc Natl Acad Sci USA 80:6177–6181
Nass K, Frenkel GD (1980) Adenovirus-specific DNA-binding protein inhibits hydrolysis of DNA
    by DNase in vitro. J Virol 35:314–319
Nevins JR, Jensen-Winkler J (1980) Regulation of early adenovirus transcription: a protein product
    from early region 2 specifically represses region 4 transcription. Proc Natl Acad Sci USA
    77:1893–1987
Ostrove JM, Rosenfeld P, Williams J, Kelly TJ Jr (1983) in vitro complementation as an assay
    for purification of adenovirus DNA replication proteins. Proc Natl Acad Sci USA 80:935–939
Pearson GD, Hanawalt PC (1971) Isolation of DNA replication complexes from uninfected and
    adenovirus infected HeLa cells. J Mol Biol 62:65–80
Petterson U (1973) Some unusual properties of replicating adenovirus type 2 DNA. J Mol Biol
    81:521–527
Pincus S, Robertson W, Rekosh D (1981) Characterization of the effect of aphidicolin on adenovirus
    DNA replication: evidence in support of a protein primer model of initiation. Nucleic Acids
    Res 19:1919–1938
Prage L, Pettersson U (1971) Structural proteins of adenoviruses. VII. Purification and properties
    of an arginine-rich core protein from adenovirus type 2 and type 3. Virology 45:364–373
Reiter T, Fütterer J, Weingärtner A, Winnacker E-L (1980) Initiation of adenovirus DNA replication.
    J Virol 35:662–671
Rekosh DMK, Russel WC, Bellett AJD, Robinson AJ (1977) Identification of a protein linked
    to the ends of adenovirus DNA. Cell 11:283–295
Revet B, Benichou D (1981) Electron microscopy of Ad5 replicating molecules after in vivo photo-
    crosslinking with trioxsalen. Virology 114:60–70
Rijnders AWM, van Bergen BGM, van der Vliet PC, Sussenbach IS (1983a) Immunological charac-
    terization of the role of adenovirus terminal protein in viral DNA replication. Virology
    131:287–295
Rijnders AWM, van Bergen BGM, van der Vliet PC, Sussenbach JS (1983b) Specific binding of
    the adenovirus terminal protein precursor-DNA polymerase complex to the origin of DNA repli-
    cation. Nucleic Acids Res 11:8777–8789
Robin J, Bourgaux-Ramoisy D, Bourgaux P (1973) Single stranded regions in replicating DNA
    of adenovirus type 2. J Gen Virol 20:233–237
Robinson AJ, Bellett AJD (1974) A circular DNA-protein complex from adenoviruses and its possible
    role in DNA replication. Cold Spring Harbour Symp Quant Biol 39:523–530
Robinson AJ, Younghusband HB, Bellett AJD (1973) A circular DNA-protein complex from adeno-
    viruses. Virology 56:54–69
Robinson AJ, Bodnar JW, Coombs DH, Pearson GD (1979) Replicating adenovirus 2 DNA mole-
    cules contain terminal protein. Virology 96:143–158
Ruben M, Bacchetti S, Graham F (1983) Covalently closed circles of adenovirus 5 DNA. Nature
    301:172–174
Schechter NM, Davies W, Anderson CW (1980) Adenovirus coded deoxyribonucleic acid binding
    protein. Isolation, physical properties, and effects of proteolytic digestion. Biochemistry
    19:2802–2810
Schick J, Baczko K, Fanning E, Groneberg J, Burger H, Doerfler W (1976) Intracellular forms
    of adenovirus DNA: integrated form of adenovirus DNA appears early in productive infection.
    Proc Natl Acad Sci USA 73:1043–1047
Schilling R, Weingärtner B, Winnacker E-L (1975) Adenovirus type 2 DNA replication. II. Termini
    of DNA replication. J Virol 16:767–774

Seif I, Khoury G, Dhar R (1979) The genome of human papovavirus BKV, Cell 18:963–977

Sergeant A, Tigges MA, Raskas HJ (1979) Nucleosome-like structural subunits of intranuclear parental adenovirus type 2 DNA. J Virol 29:888–898

Sharp PA, Moore C, Haverty JC (1976) The infectivity of adenovirus 5 DNA-protein complex. Virology 75:442–458

Shaw CH, Rekosh DMK, Russell WC (1980) Catalysis of adenovirus DNA synthesis in vitro by DNA Polymerase γ. J Gen Virol 48:231–236

Shimojo H, Yamashita T (1968) Induction of DNA synthesis by adenoviruses in contact-inhibited hamster cells. Virology 36:422–433

Shinagawa M, Padmanabhan R (1980) Comparative sequence analysis of the inverted terminal repetitions from different adenoviruses. Proc Natl Acad Sci USA 77:3831–3835

Smart JE, Stillman BW (1982) Adenovirus terminal protein precursor. J Biol Chem 257:13499–13506

Steenbergh PH, Maat J, van Ormondt H, Sussenbach JS (1977) The nucleotide sequence at the terminus of adenovirus type 5 DNA. Nucleic Acids Res 4:4371–4389

Stillman BW, Bellett AJD (1979) An adenovirus protein associated with the ends of replicating DNA molecules. Virology 93:69–79

Stillman BW, Bellett AJD, Robinson AJ (1977) Replication of linear adenovirus DNA is not hairpin-primed. Nature 269:723–725

Stillman BW, Lewis JB, Chow LT, Mathews MB, Smart JE (1981) Identification of the gene and mRNA for the adenovirus terminal protein precursor. Cell 23:497–508

Stillman BW, Topp WC, Engler JA (1982a) Conserved sequences at the origin of adenovirus DNA replication. J Virol 44:530–537

Stillman BW, Tamanoi F, Mathews MB (1982b) Purification of an adenovirus-coded DNA polymerase that is required for initiation of DNA replication. Cell 31:613–623

Stillman BW (1983) The replication of Adenovirus DNA with purified proteins. Cell 35:7–9

Subramanian PH, Maat J, van Ormondt H, Sussenbach JS (1977) Nucleotide sequence of a fragment of SV40 DNA that contains the origin of DNA replication and specifies the 5′-ends of "early" and "late" viral RNA. J Biol Chem 252:355–367

Sugisaki H, Sugimoto K, Takanami M, Shiroki K, Saito Y, Shimojo Y, Sawada Y, Uemizu Y, Uesigi S-I, Fujinaga K (1980) Structure and gene organization of the transforming HindIII-G fragment of Ad12. Cell 20:777–786

Sung MT, Cao TM, Coleman RT, Budelier KA (1983) Gene and protein sequences of adenovirus protein VII a hybrid basic chromosomal protein. Proc Natl Acad Sci USA 80:2902–2906

Sussenbach JS, Kuijk MG (1977) Studies on the mechanism of replication of adenovirus DNA. V. The location of termini of replication. Virology 77:149–157

Sussenbach JS, Kuijk MG (1978a) The mechanism of replication of adenovirus DNA. VI. Localization of the origins of the displacement synthesis. Virology 84:509–517

Sussenbach JS, Kuijk MG (1978b) Initiation of adenovirus DNA replication does not occur via a hairpin mechanism. Nucleic Acids Res 5:1289–1295

Sussenbach JS, van der Vliet PC (1972) Viral DNA synthesis in isolated nuclei from adenovirus infected KB cells. FEBS Lett 21:7–15

Sussenbach JS, van der Vliet PC (1982) In: Becker Y (ed) Molecular events in the replication of viral and cellular genomes. Adenovirus DNA replication. Developments in molecular biology 2. Nijhoff, The Hague

Tamanoi F, Stillman BW (1982) Function of the adenovirus terminal protein in the initiation of DNA replication Proc Natl Acad Sci USA 79:2221–2225

Tamanoi F, Stillman BW (1983) Initiation of adenovirus DNA replication in vitro requires a specific DNA sequence. Proc Natl Sci USA 80:6446–6450

Tate VE, Philipson L (1979) Parental adenovirus DNA accumulates in nucleosome-like structures in infected cells. Nucleic Acids Res 6:2769–2785

Temple M, Antoine G, Delius H, Stahl S, Winnacker E-L (1981) Replication of mouse adenovirus strain Fl DNA. Virology 109:1–12

Tolun A, Pettersson U (1975) Termination sites for adenovirus type 2 DNA replication. J Virol 16:759–766

Tolun A, Aleström P, Pettersson U (1979) Sequence of inverted terminal repetitions from different adenoviruses: demonstration of conserved sequences and homology between SA7 termini and SV40 DNA. Cell 17:705–713

Tomizawa J, Ogawa T (1968) Replication of phase lambda DNA. Cold Spring Harbor Symp Quant Biol 33:533–551

Tyndall T, Younghusband HB, Bellett AJD (1978) Some adenovirus DNA is associated with the DNA of permissive cells during productive or restricted growth. J Virol 25:1–10

Van Bergen GM, van der Vliet PC (1983) Temperature-sensitive initiation and elongation of adenovirus DNA replication in vitro with nuclear extracts from H5ts36-, H5ts149-, and H5ts125-infected HeLa cells. J Virol 46:642–648

Van Bergen BGM, van der Ley PA, van Driel W, van Mansfeld ADM, van der Vliet PC (1983) Replication of origin containing adenovirus DNA fragments that do not carry the terminal protein. Nucleic Acids Res 11:1975–1989

Van der Eb AJ (1973) Intermediates in type 5 adenovirus DNA replication. Virology 51:11–23

Van der Vliet PC, Levine AJ (1973) DNA-binding proteins specific for cells infected by adenovirus. Nature 246:170–173

Van der Vliet PC, Sussenbach JS (1972) The mechanism of adenovirus DNA synthesis in isolated nuclei. Eur J Biochem 30:548–592

Van der Vliet PC, Sussenbach JS (1975) An adenovirus type 5 gene function required for initiation of viral DNA replication. Virology 67:415–426

Van der Vliet PC, Levine AJ, Ensinger MJ, Ginsberg HS (1975) Thermolabile DNA-binding proteins from cells infected with a temperature-sensitive mutant of adenovirus defective in viral DNA synthesis. J Virol 15:348–354

Van der Vliet PC, Zandberg J, Jansz HS (1977) Evidence for a function of the adenovirus DNA-binding protein in initiation as well as in elongation of nascent DNA chains. Virology 80:98–110

Van der Werf S, Bouché J-P, Michali M, Girard M (1980) Involvement of DNA polymerases α and γ in replication of adenovirus deoxyribonucleic acid in vitro. Virology 104:56–72

Van Wielink PS, Naaktgeboren N, Sussenbach JS (1979) Presence of protein at the termini of intracellular adenovirus type 5 DNA. Biochim Biophys Acta 563:89–99

Watson JD (1972) Origin of concatemeric T7 DNA. Nature 239:197–201

Weingärtner B, Winnacker E-L, Tolun A, Pettersson U (1976) Two complementary strand-specific termination sites for adenovirus DNA replication. Cell 9:259–268

Weissbach A (1975) Vertebrate DNA polymerases. Cell 5:101–108

Wilkie NM, Ustacelibi S, Williams F (1973) Characterization of temperature-sensitive mutants of a adenovirus type 5 nucleic acid synthesis. Virology 51:499–503

Williams FF, Young H, Austin P (1974) Genetic analysis of adenovirus type 5 in permissive and nonpermissive cells. Cold Spring Harbor Symp Quant Biol 39:427–437

Williams J, Galos RS, Binger MH, Flint SJ (1979) Location of additional early regions within the left quarter of the adenovirus genome. Cold Spring Harbour Symp Quant Biol 44:353–365

Winnacker E-L (1975) Adenovirus type 2 DNA replication. I. Evidence for discontinuous DNA synthesis. J Virol 15:744–758

Winnacker E-L (1978) Adenovirus DNA: Structure and function of a novel replicon. Cell 14:761–773

Wist E, Prydz H (1979) The effect of aphidicolin on DNA synthesis in isolated HeLa cell nuclei. Nucleic Acids Res 6:1583–1590

Yamashita T, Arens M, Green M (1975) Adenovirus deoxyribonucleic acid replication. II. Synthesis of viral deoxyribonucleic acid in vitro by a nuclear membrane fraction form infected KB cells. J Biol Chem 250:3273–3279

A Study of Viral Genomes in Cells Transformed by the Nononcogenic Human Adenovirus Type 5 and Highly Oncogenic Bovine Adenovirus Type 3

E.I. FROLOVA and E.S. ZALMANZON

1 Introduction

Over 20 years ago HUEBNER et al. (1962) published the first paper on the ability of Ad12 to induce tumors in hamsters. This was followed by publication of a mass of data characterizing both the different biological properties of cells transformed by adenoviruses and tumor cells, and the genomes of adenoviruses and the cells transformed by them.

It was shown that almost all human and some animal adenoviruses can transform rodent cells in vitro, but that not all such in vitro-transformed cells can induce tumors in the respective hosts.

It is still not known what determines the oncogenicity of in vitro-transformed cells: – the organism of the animal, its immunological reactivity, properties of the proteins encoded by the viral oncogene, or rearrangements of the genome of transformed cells, which may depend on the sites of integration of viral genome into the genome of the cell.

Nor is it clear whether cells transformed in vitro themselves have properties that determine their oncogenicity upon injection to animals or whether the oncogenic properties depend solely on the virus that induced their transformation.

DNA of rat and hamster cells transformed by group C adenoviruses contains from 6% of the viral genome up to all or almost all of it, and in different

Institute of Molecular Biology, The Academy of Sciences of the USSR, Moscow, USSR cell lines there are from 1.6 to 13.3 copies of it per diploid amount of cellular DNA (GALLIMORE 1974; SAMBROOK et al. 1974; SHARP et al. 1974a, b).

Rat and hamster cells transformed by group A and B adenoviruses contain from 50% up to all or almost all of the viral genome in large numbers of copies (FANNING and DOERFLER 1976; GREEN et al. 1977; LEE and MAK 1977; STABEL et al. 1980).

It might be assumed that the size and the number of copies of virus-specific sequences inserted into the cellular DNA determine its oncogenic properties. But KUHLMAN et al. (1982) have described cells of hamster tumors induced by Ad12 which contain one or two copies of the entire viral DNA molecule or its terminal fragments. In other experiments cell transformation was induced by restriction fragments of DNA of oncogenic viruses Ad7, Ad12, and BAV3. Cells transformed by adenoviral DNA fragments that comprise from 7.2% to 11.9% of its length and are located at its left end have the same transforming activity as cells transformed by the whole viral DNA molecule (SHIROKI et al. 1977, 1979; IGARASHI et al. 1978; DIJKEME et al. 1979).

BERNARDS et al. (1981, 1982) have constructed hybrid plasmids which contain region E1A (1.3–4.5 map units) of Ad5 and region E1B (4.5–11.2 map units) of Ad12, and vice versa. They transformed the primary kidney cell culture of newborn rats by these plasmids and showed that the oncogenicity was determined by region E1B of Ad12. VAN DER ELSEN et al. (1981) have shown that region E1A is necessary for the beginning of the transformation, and E1B, for complete transformation.

According to BYRD et al. (1982), plasmids that contain 6.8% and 4.7% of the map units of Ad12 DNA have significantly less than complete transforming activity. In experiments conducted by JOCHEMSEN et al. (1980, 1982) none of the cell lines transformed by the HindIII-G fragment of Ad12 DNA (0–7.2 map units) was able to induce tumors in nude mice and hamsters, although these cells were transformed morphologically and contained T antigens. Cells transformed by the EcoRI-C fragment (0–16 map units) were tumorigenic. NIIYAMA et al. (1981) transformed a cloned mouse BALB 3T3-A31 cell line with a whole BAV3 viral DNA molecule, with its EcoRI-D fragment (3.6–19.4 map units), and with ligated fragments HindIII-J and -E (0–11.9 map units). While the oncogenicity of cells transformed by the whole viral DNA molecule and by its EcoRI-D fragment were the same, the oncogenicity of cells transformed by the HindIII-J and -E fragments was lower by two orders of magnitude.

It may therefore be considered established that the transforming and oncogenic activities of different groups of adenoviruses are determined by the properties of their oncogenic fragments (oncogenes), which have distinctive primary structures (BRUSKA et al. 1981). DNAs of all transformed cells contain some definite sequences of the viral genome, which are responsible for the initiation of transformation and its maintenance irrespective of whether the transforming agent is the whole virus, its DNA, or a specific fragment of this DNA.

The mechanism of adenoviral DNA integration into the cellular DNA and the specificity of integration sites deserve special attention.

Investigations carried out by several groups of scientists in a model of the oncogenic virus Ad12 and nononcogenic viruses Ad2 and Ad5 have shown that DNAs of different cell lines transformed by these viruses in vitro and DNAs of tumor cells differ both in the number and in the size of the fragments carrying virus-specific nucleotide sequences (FANNING et al. 1978; FROLOVA and GEORGIEV 1979a; DOERFLER et al. 1979; IBELGAUFTS et al. 1980; SAMBROOK et al. 1979; STABEL et al. 1980; VISSER et al. 1979; DORSCH-HÄSLER et al. 1980; VARDI-MON and DOERFLER 1981; DOERFLER 1982; EICK and DOERFLER 1982).

The authors conclude that viral DNA and its terminal fragments are inserted into different repetitive nucleotide sequences of cellular DNA and that right and left ends of the viral DNA molecule can be inserted together (the possibility that in some cases they are separated by short pieces of cellular DNA is not excluded (DOERFLER 1982; EICK and DOERFLER 1982).

The existence of multiple sites of integration of viral DNA and its fragments into the genome of the cell is indicative of their accidental insertion, but it does not rule out the possibility that there are multiple unique sites and that integration into these leads to the cell transformation.

This assumption is supported by results of DEURING et al. 1981a, b), GAHL-MAN et al. (1982), and DOERFLER (1982), who cloned left-terminal fragments of Ad2 and Ad12 DNAs with adjacent cellular sequences, determined their primary structure, and found short regions of homology between cellular and viral sequences 8–11 base pairs long as inverted repeats at the sites of integration.

It was shown that mouse cells transformed by SV40 and polyoma also contain inverted cellular nucleotide sequences adjacent to the viral sequences (MUR-PHY et al., personal communication; BOURGAUX 1982).

BOTCHAN et al. (1976), KETNER and KELLY (1976), WOLD et al. (1980), STRIN-GER (1981), and RULEY et al. (1982), however, have shown that DNAs of SV40 and polyoma virus are accidentally inserted into cellular DNA. Different lines of transformed cells have different sites of integration and as a rule they have no insertion sites in common.

It has also been shown that in the genome of nontransformed cells there are sequences complementary to DNA of Ad2, Ad5, and Ad6 (FROLOVA and GEORGIEV 1979b; JONES et al. 1979; TIKHONENKO et al. 1981a). It is possible that these regions determine the sites of insertion of viral genomes.

Some of the data available indicate functional heterogeneity of identical fragments of viral DNA inserted into cellular DNA. FLINT and SHARP (1976), FLINT and WEINTRAUB (1977), FROLOVA et al. (1977, 1978), and FROLOVA and ZALMANZON (1978) have shown that not all identical sequences of viral DNA inserted into cellular DNA are transcribed in the cells transformed by Ad2 and Ad5. However, these investigations cannot be regarded as complete.

The question as to whether there are specific sites of integration of viral oncogene into cellular DNA is closely connected with the problem of the mechanism of this phenomenon. In the past articles have appeared showing that the integration of polyoma virus (DELLA VALLE et al. 1981; NEER et al. 1983) and Ad12 (DOERFLER 1982) can be accompanied by deletion of an adjacent piece of cellular DNA and of terminal sequences of viral DNA. These data can probably prove that the integration of viral DNA is based on recombination (BOTCHAN et al. 1980). MOUGREAU et al. (1980) have also suggested the possible existence of definite sites of recombination. There are questions that have not yet been solved: Does viral DNA integrate first into single sites of cellular

DNA, which are then amplified together with virus-specific sequences, or are fragments of viral DNA inserted simultaneously into different sites of cellular DNA? Does the oncogenicity of in vitro-transformed cells depend on the viral oncogene amplification? How do the inserted tranforming regions of viral DNA affect the properties of the cells? What characteristics of in vitro-transformed cells correlate with their oncogenicity for animals? What are the normal cellular nucleotide sequences complementary to the virus-specific ones?

Some of these questions have been studied during the past few years in G. Georgiev's laboratory in the Institute of Molecular Biology, Academy of Sciences of the USSR.

2 Biological Properties of Rat Embryo Cells Transformed In Vitro by Human Ad5 and Bovine Ad3 DNA

Nine lines of rat embryo cells transformed by human Ad5 and by its DNA, two lines of rat embryo cells transformed by BAV3 DNA, and one line of hamster embryo fibroblasts transformed by BAV3 virus were obtained. A number of properties of these cell lines were studied: cell morphology by light and scanning electron microscopy, their ability to grow in soft agar, dependence on serum factors, karyotype, fibronectin distribution, and oncogenicity. In addition, the size and the number of copies of virus-specific sequences in the DNA of transformed cells, as reflected in the change of the renaturation rate of highly labeled fragments of viral DNA in the presence of DNA of transformed cells, were studied (ZALMANZON et al. 1979, 1981, 1982). The WBR1 strain of BAV3 virus had been cloned twice in advance by the plaque technique.

The biological properties of some of the cell lines studied are given in Table 1.

The cells transformed by Ad5 DNA had an epithelioid-type morphology and had a tendency to multilayer growth. Their growth was independent of the serum factors; about 7–8% of the cells produced large colonies in soft agar. Fibronectin was practically absent from the cells and they differed from normal fibroblasts in having a great number of microfibrils and ruffles on their surface. Even when syngeneic newborn rats were treated with antilymphocyte serum, no tumors were found during the subsequent 11-month observation period.

The cells transformed by BAV3 DNA differed less markedly from normal fibroblasts: they had fibroblast-like morphology and grew in monolayers; they had only isolated microfibrils and ruffles on scanning electron microscopy; and their dependence on serum factors was more prominent. Colonies of such cells in soft agar were small and the number of colonies was lower. They contained less fibronectin than normal fibroblasts and fibronectin was visible on the cell border.

CHEN et al. (1976) had observed the same in cells transformed by Ad2, but the cells transformed by Ad5 DNA contained no fibronectin at all. The rat embryo cells transformed by BAV3 DNA induced tumors in immunocompetent newborn rats (fibroblastic sarcomas) but the cells transformed by Ad5 DNA did not.

Table 1. Properties of rat embryo cells transformed in vitro by DNAs of adenoviruses type 5 and BAV3

Cell line	DNA source	Cell morphology	Effect of serum concentration (cell no. after 90 h: cell no after 40 h)		Percentage of colonies in soft agar	Modal number of chromosomes	Fibronectin content	Type specificity T antigen	Oncogenicity			
			1-2% Calf serum	10% Calf serum					Immunol. reactivity	No. of cells inoculated	Latency period (weeks)	No. of tumors obtained/No. of rats inoculated
FK$_{Aug}$	0	Fibroblastoid	1.0	2.6	0[b]	40–50	+ + + +	0	–	–	–	–
DNA-2A5	Ad5	Epithelioid	3.2	2.8	6.7	20–50	0	–	immuno-depr.	2.0 × 10^6	?	0/7(11)[c]
DNA-3A5	Ad5	Epithelioid	–[a]	–	8.0	70–80	–	+		2.2 × 10^6	?	0/6(11)[c]
ADB2a/4	BAV3	Fibroblastoid	1.5	5.2	1.4	20–50	+	+	immuno-comp.	2.0 × 10^6	8	7/8
ADB4	BAV3	Fibroblastoid	1.5	3.7	1.1	40–50	+	+		6.0 × 10^5	8	2/4

a   Not studied
b   Negative results
c   Sacrificed after 11 months

Thus, the cells transformed by nononcogenic Ad5 turned out to be more markedly transformed with reference to all the properties studied than the cells transformed by BAV3 DNA, but they did not induce tumors. Our results support the earlier data published by STILES et al. (1976) showing the absence of any correlation between properties of in vitro-transformed cells and their ability to induce tumors in nude mice and the data published by GALLIMORE et al. (1979), GALLIMORE and PARASKEVA (1980) and PARASKEVA and GALLIMORE (1980) showing the absence of any correlation between the ability of different lines of Ad12-transformed cells to form colonies in methylcellulose, the fibronectin content, and their oncogenicity.

Rat embryo cells transformed by BAV3 DNA did not differ from hamster embryo cells transformed by the whole virus in their oncogenicity for hosts.

Our results confirm the leading role of the viral oncogene in tumor formation.

3 Investigation of the Cells Transformed by BAV3 DNA Before and After Passage Through the Host Organism

It is known that passage through the host organism increases the oncogenicity of in vitro-transformed cells (STRIZHACHENKO et al. 1974; HARWOOD and GALLIMORE 1975; ADELMAN et al. 1981).

We used the study of transformed cells before and after such a passage to investigate the question of what properties determine the oncogenicity of in vitro transformed cells and might be correlated with it.

We obtained cloned cell cultures from four rat tumors after SC injection of ADB2a/4 and ADB4 cells (cell lines OAV2a/4, Op2a/4, and OAV4, OAV4a/1, and OAV4a/2) and from hamster tumors after injection of hamster embryo cells transformed by BAV3 (lines HE4 and HE40). The passage of in vitro-transformed cells through the host organism resulted in an increase of their oncogenicity: the length of the latent period was reduced from 8 weeks to 2–3 weeks; the number of cells required to induce tumors was lower; and it was possible to induce tumors not only in syngeneic but also in mongrel rats. This did not affect the cell morphology (as seen under the light and scanning electron microscopes, or their karyotype, dependence of their growth on serum factors, or the percentage of colonies formed in soft agar.

The fibronectin content did not change in one of the cell lines studied, but it decreased in the second. This change did not correlate with the oncogenicity of the cells, however. Only a slight parallelism was observed between the proliferative activity of the cells in the medium with 10% serum and their oncogenicity.

4 Sialyltransferase Activity in the Cells Transformed by BAV3 DNA Before and After Passage Through the Host Organism

Many authors connect the oncogenic properties of transformed cells with the changes in their membranes.

Table 2. Comparison of activity of membrane-bound sialyltransferases and the content of sialic acids in rat embryo cells transformed by BAV3 DNA before and after a passage in the animal

Cell line	Radioactivity $(cmp \times mg^{-1}$ protein)		Specific activity $(pmol \times mg^{-1}$ protein)		Exogenous/ endogenous	Content of sialic acids $(nmol \times mg^{-1}$ protein)
	Exogenous	Endogenous	Exogenous	Endogenous		
Normal fibroblasts	3900 ± 800	1400 ± 300	5.86	2.10	2.79	22.3
ADB2a/4	6500 ± 650	1350 ± 320	9.76	2.03	4.81	20.5
Op2a/4	8440 ± 600	1130 ± 340	12.67	1.70	7.45	27.2
OAV2a	6100 ± 580	470 ± 80	9.16	0.70	13.10	16.8
ADB4	16250 ± 1000	3750 ± 200	24.40	5.63	4.33	22.8
OAV4	<200	3500 ± 300	–	5.50	–	12.8
OAV4a/2	8480 ± 650	4200 ± 500	12.73	6.31	2.02	–

–, not studied

Changes in the content and properties of glycoproteins, glycolipids, and sialic acids have been described which correlate with their oncogenicity (VAN BEEK et al. 1973, 1975; RICHARDSON et al. 1975; FRANKE 1977; WARREN et al. 1978; BRAMWELL and HARRIS 1978, 1979).

It has been shown that in highly metastatic mouse cells transformed by the Kirsten Sarcoma virus, sialization of glycoproteins and glycosphingolipids is more pronounced than in weakly metastatic cells transformed by the same virus (YOGESWARAN et al. 1979).

AOI and JAKOTA (1978) have studied glycoproteins and sialyltransferase activity in the early stage of infection of human embryo cells with variants of Ad12 that differed in their oncogenicity. More oncogenic variants of the virus and less oncogenic ones induced the synthesis of different glycoproteins and stimulated sialyltransferase activity in microsomes. The authors showed a good correlation between sialyltransferase activation and oncogenicity. According to LLOYD (1975), sialic acids "overprevent" recognition of a cell by immunocompetent cells, and an excess of these ensures the absence of contact inhibition in transformed cells.

In collaboration with GABRIELYAN et al. (1980, 1984), we have studied the activity and multiple forms of sialyltransferases in rat embryo cells transformed by BAV3 DNA before and after passage of these cells through the host organism.

We used the asialoglycoproteins fetuin and mucine from pig submaxillar glands and a low-molecular-weight N-acetyllactose amine as substrates for the determination of exogenic sialyltransferase activity. Endogenic enzyme activity was determined without substrate addition. We determined the incorporation of CMP-[^{14}C]NANA into the total membrane fraction of the cellular extract of a 4-day-old culture treated with neuraminidase; multiple enzyme forms were studied by electrofocusing in sucrose density gradient over the pH range 3.0–10.0. The optimum pH was also determined by comparison of enzyme activities at pH levels ranging from 5.2 to 8.5. The results are given in Table 2.

Different cell lines obviously varied in sialic acid content, but the difference was coincidental. Increased exogenous sialyltransferase activity was observed in all lines of transformed and tumor cells (with the exception of line OAV4, in which the enzyme that would have affected the substrates used was practically absent). No difference was found in enzyme activities between in vitro-transformed and tumor cells. In one group of cells an increase of exogenic sialyltransferase activity was accompanied by unchanged (lines ADB2a and Op2a/4) or lowered endogenic enzyme activity (line OAV2a). A significant activation of enzyme by endogenous substrates was observed in the second group. Different cell lines contained multiple enzyme forms with different pH optima.

The results obtained support data of other authors on changes in cellular glycoproteins upon transformation, which is revealed in activation and change of the sialyltransferase pattern. But no parallelism was observed between enzyme activation and the degree of the increase in oncogenicity upon passage through the host organism.

5   Integration of the BAV3 Genome into DNA of In Vitro-Transformed Cells and Tumor Cells in Culture

The next question, which has to be studied very carefully, is concerned with the amplification of the viral oncogene and whether this takes place in the host organism.

Up to now different authors have studied either in vitro-transformed cells or cells taken from tumors induced by viruses. There are few data on the comparison of the genomes of transformed cells before and after their passage through the host organism, and these data are rather contradictory.

FRENKEL et al. (1976) compared the line of cells transformed by herpesvirus 2 after 80 passages in vitro with the line of tumor cells obtained from that line. They showed that tumor cells in culture are simpler than those in original culture, but the number of copies of such sequences is greater (0.7 of a copy from 16% of viral DNA length in original culture and 3 copies from 8% of genome in the culture of tumor cells).

RESSONS et al. (1979), however, found 6.0 and 4.3 copies from 43% and 39% of the viral DNA molecule, respectively, in two cell lines transformed in vitro by herpesvirus 2 before injection to animals and 2.2 copies from 47% and 43% of the viral DNA molecule in cultures of tumor cells.

VAN DER PUTTEN et al. (1979) discovered the amplification and reintegration of virus-specific sequences during leukemogenesis by Moloney virus. But VARDIMON and DOERFLER (1981) did not find any change in virus-specific sequences in a hamster tumor in comparison with the original cells transformed by Ad12. Then MORIUCHI et al. (1982) showed that cells of tumors which developed in adult rats after injection of rat cells transformed by a HindIII-G fragment of Ad12 DNA (0–6.8 map units) contained single, but not multiple sites of integration into the host DNA.

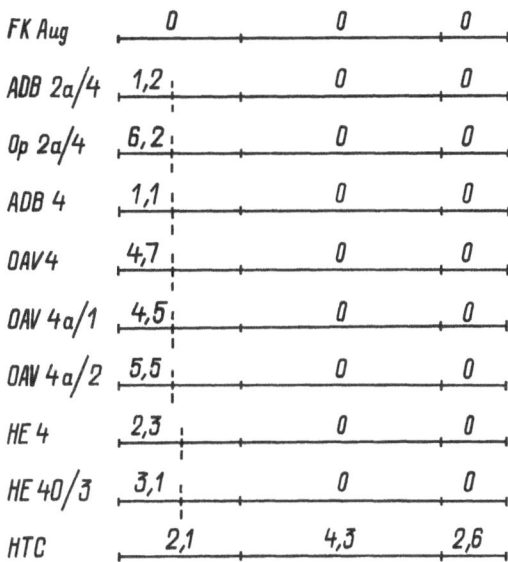

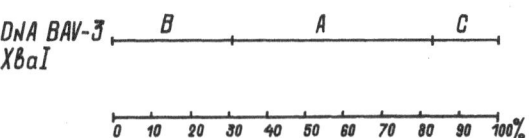

Fig. 1. The content of sequences complementary to *Xba*I fragments of BAV3 DNA, in DNA of in vitro transformed and tumor cells (a scheme). *Numbers* represent the number of viral DNA copies per diploid amount of cellular DNA

TICHONENKO et al. (1981 b, c) did not find qualitative or quantitative changes of viral sequences after passage through the host organism of rat cells transformed by a mixture of *Sal*I fragments of SA7 DNA. According to LANIA et al. (1980), tumor cells obtained after injection to animals of cells transformed by polyoma virus contained a lowered number of virus-specific sequences and, in contrast to the original culture, did not contain a free viral genome.

We determined the renaturation kinetics of highly labeled viral DNA fragments in the presence of the DNA from the transformed and tumor cells. We used the fragments produced when BAV3 DNA was restricted with *Xba*I and labeled with [32]P.

DNA from normal rat embryo fibroblasts and from HTC cells (obtained from N.M. STRIZHACHENKO; cells of primary hamster tumor that had undergone 23 passages in vivo and more 1000 passages in vitro) (GRAEVSKAYA et al. 1972) were used as controls. DNA of calf thymus was used as a carrier (ZALMANZON et al. 1982). We found 1.2 and 1.1 copies from 13.2% of the left end of BAV3 DNA in in vitro-transformed rat cells and 2.3 copies from 16.5% of viral DNA molecule in hamster cells, calculated on the basis of the diploid amount of cellular DNA.

Passage through the host organism was accompanied with an increase to 4.5–6.2 in the number of oncogene copies in rat cells and to 3.1 in hamster cells. The size of inserted fragments was virtually unchanged. HTC cells con-

Table 3. Content of virus-specific sequences in DNA of in vitro transformed and tumor cells and their tumorigenicity

Cell line	Characteristics of cells	No. of oncogene equivalents per diploid cell genome	TD_{50} (for syngeneic rats)
ADB2a/4	Transformed in vitro	1.2	$10^{5.5}$ [a]
Op2a/4	Tumor	6.2	$10^{3.1}$ [b]
OAV2a	Tumor	–	$10^{2.2}$ [b]
ADB4	Transformed in vitro	1.1	$10^{5.5}$ [a]
OAV4	Tumor	4.7	$10^{4.5}$ [b]
OAV4a/1	Tumor	4.5	–
OAV4a/2	Tumor	5.5	$10^{4.3}$ [b]

–, not studied
[a]   4 months after inoculation
[b]   1.5 months after inoculation

tained the whole viral genome, but its middle part was represented by a greater number of copies (4.3) than the terminal parts (B fragment, 2.1 copies; C fragment, 2.6 copies).

The large DNA fragments used in the renaturation kinetics experiments do not allow exclusion of the existence of deletions of virus-specific sequences in the DNA of any of the cell lines studied, especially of HTC cells. The results of experiments are given in Fig. 1.

Thus the experiments performed have shown that passage through the animal organism results in increased oncogenicity of in vitro-transformed cells, accompanied by an increase in the number of oncogene copies in their genomes. To find out whether there is a direct correlation between these two events experiments were carried out to determine the tumorigenicity of the cells following SC inoculation. The results are given in Table 3.

The number of oncogene copies was found to be increased approximately five-fold in cell lines Op2a/4 and OAV4a/2 in comparison with the cells of the original cultures, while they differed in their oncogenicity by two orders of magnitude. At the same time HTC cells, which induce tumors at a dose of 10^1 (STRIZHACHENKO et al. 1974), according to our results, contain 2.1 oncogene copies per diploid amount of cellular DNA. Thus no direct correlation has been found.

6   Distribution of Viral Genes Concerning DNA-Protein Axial Structure and DNA Loops of Histone-Depleted Metaphase Chromosomes

Investigation of the specificity of integration of viral genomes in cell chromosomes is highly significant for the understanding of viral transformation mechanisms. Several mechanisms can be suggested for specific integration of viral sequences during transformation. Probably a special set of regions related to cell transformation exists in the DNA of normal cells; that is to say, integration into these regions is presumably accompanied by cell transformation. On the other hand, there may be specific host DNA sequences that function as targets for the integration of foreign DNA.

In any case, investigation of this question requires that cell genomic regions bordering on inserted sequences be analyzed in a wide range of integration events, and their organization compared. Such experiments are usually based on analysis of the restriction endonuclease pattern of adjacent regions by the blotting technique and determination of the nucleotide sequences of such regions.

Certainly there may be higher levels of integration specificity, e.g., localization of integrated viral genes in the unique host chromosome or in specific parts of metaphase or interphase chromosomes.

The most obvious demonstration of the existence of axial structures and attached DNA loops was obtained by Laemmli and co-workers by electron-microscopic investigation of histone-depleted metaphase chromosomes (PAULSON and LAEMMLI 1977; ALOLPH et al. 1977). Metaphase chromosomes had the sedimentation coefficient 4000S–7000S (after extraction of histones with 2 M NaCl or with a mixture of such polyanions as heparin and dextran sulfate). On the electromicrographs such chromosomes were visualized as axial structures surrounded by supercoiled DNA, and two chromatids could be identified. The length of the axial structure was equal to the length of the original chromosome. Such histone-depleted chromosomes were dissociated by treatment with 4 M urea, 0.1% SDS, or proteases. DNAseI almost completely eliminated supercoiled DNA loops, but axial structures were resistant to DNAseI. In such chromosomes the length of the DNA loops was 50000–100000 base pairs (PAULSON and LAEMMLI 1977).

We have investigated the distribution of viral DNA sequences concerning DNA protein axial structures and DNA loops of histone-depleted metaphase chromosomes of Ad5-transformed cells (E.I. FROLOVA, V.L. METT, E.S. ZAL-MANZON, G.P. GEORGIEV, unpublished data). For such experiments we selected the line of rat embryo cells transformed by Ad5 that contains approximately 15% of viral left-end sequences, line FK Ad5 (FROLOVA and ZALMANZON 1978). Histone-depleted metaphase chromosomes were isolated from the synchronized cells by a modification (RAZIN et al. 1978) of the method of LAEMMLI et al. (1978). The axial structure DNP was separated from the loop DNA of histone-depleted chromosomes by digestion with *Bam*HI restriction endonuclease. Approximately 7% of the total chromosomal DNA remains stably bound to axial DNP structures. Equal amounts of DNAs extracted from loops and axial structures separated by sucrose gradient centrifugation were then completely digested by the same endonuclease and the integration pattern of viral DNA was analyzed by Southern blotting (SOUTHERN 1975) and hybridization using ^{32}P-labeled Ad5 DNA (WAHL et al. 1979). Two viral DNA-specific bands 6.5 and 2.1 kb in length appeared in *Bam*HI-restricted DNA of loops (Fig. 2). A very slow viral DNA-specific band was also detected in DNA of axial structure; this was approximately 12 kb in length.

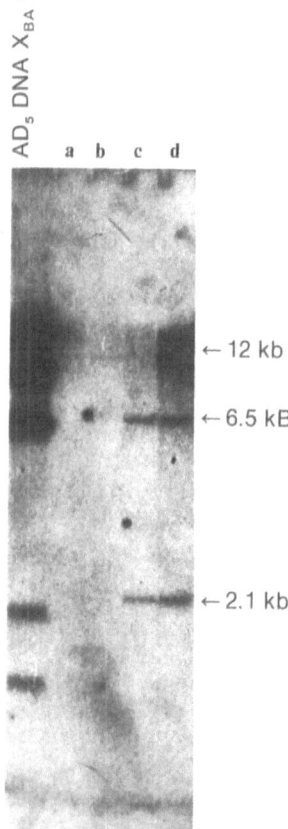

AD₅ DNA X_BA I

a   b   c   d

← 12 kb

← 6.5 kB

← 2.1 kb

Fig. 2a–d. Hybridization of viral DNA to the DNA of axial chromosome structures (**a** and **b**) and DNA of loops (**c** and **d**). **a, c** DNA completely digested by *Bam*HI after separation of axial structures and loops

Thus most viral DNA sequences of line FKAd5 were localized in the DNA loops of histone-depleted chromosomes, and only a minority could be detected in a high-molecular-weight fragment. Such asymmetrical distribution of integrated viral sequuences within the structure of metaphase chromosomes, together with the existence of limited numbers of viral-DNA-specific fragments, demonstrates the nonrandom organization of cellular DNA in the loop-axial structure in the metaphase chromosomes. It could be suggested that the majority of chromosomes bearing the given marker (integrated viral genes) had the same type of organization. This conclusion is in agreement with the data concerning the distribution of repetitive sequences, such as satellite DNA or the most abundant classes of moderate repeats (Razin et al. 1978). On the other hand, on the basis of the asymmetrical distribution of viral sequences we can propose different functional significances of different points of integration of viral genes in cellular DNA. It will be very interesting to pursue this approach to the question of viral DNA integration, to elucidate whether there is any specificity of the discovered localization of viral genes in the loops in other transformed cell lines.

7 Patterns of Adenovirus DNA Integration in Different Lines of Ad5-Transformed Cells

The main problem still to be solved by analysis of the integration pattern of viral DNA is the question of specificities at the insertion sites. The data based on the study of a large number of adenovirus-, SV-40-, and polyoma-transformed and tumor cell lines suggest that there is no specific host cell DNA sequence in which viral DNA is inserted (KETTNER and KELLY 1976; BOTCHAN et al. 1976; SUTTER et al. 1978; DOERFLER et al. 1979; SAMBROOK et al. 1979; VISSER et al. 1979; FROLOVA and GEORGIEV 1979a; STABEL et al. 1980; TICHO-NENKO et al. 1981a; KUHLMANN and DOERFLER 1982).

We have analyzed the pattern of adenovirus DNA integration in four independently isolated lines of rat embryo cells transformed by Ad5 and its DNA (FROLOVA and ZALMANZON 1978). Experimental data obtained with restriction endonucleases *Eco*RI, *Hin*dIII, and *Xba*I were used to design the most likely models describing the integration pattern of Ad5 DNA in FKAd5, DFKI, DFK3, and FKAugust DNA3 cell lines. These models are shown in Fig. 3. Viral DNA-specific bands that appeared on autoradiograms and did not co-migrate with any of the simultaneously electrophoresed marker virion DNA bands were designated off-size bands and it was suggested that they consisted of covalently linked viral and cellular DNA. We can distinguish the following most characteristic features of the integration pattern obtained for Ad5 DNA and shown in Fig. 3. We cannot detect terminal fragments of viral DNA in any case, which means that the terminal regions of Ad5 DNA were linked to cellular DNA. The integration pattern of every cell line investigated was unique and specific for each cell line over a long period of cell propagation. In several lines there were fewer sites of integration than Ad5 DNA copies (FROLOVA and GEORGIEV 1979a). This can be explained by the suggestion that viral DNA sequences were amplified upon integration together with adjacent cellular DNA. Inserted viral DNA sequences are present in transformed DNA, and in a more intact form in cells transformed by Ad5 virions (FKAd5, DFKI, and DFK3) than in cells transformed by viral DNA (FKAugust DNA3).

Our studies of the integration of the Ad5 genome into the genome in transformed rat cells led us to the conclusion that the integration pattern of the viral genome is unique for each cell line. On the other hand, in *Bam*HI-restricted DNA of each transformed cell line we found several rather short fragments in the range of 1.3–5.3 kb binding viral DNA during hybridization (FROLOVA and GEORGIEV 1979b). *Bam*HI endonuclease cuts Ad5 DNA at the only site near the middle of the molecule, so it might be expected that most of the labeled viral DNA would hybridize to high-molecular-weight fragments (more than 10 million daltons) in almost all cell lines. These long cellular DNA fragments probably contain integrated viral DNA linked covalently to host DNA sequences. However, besides hybridization to these long DNA fragments specific for each line of transformed cells in all transformed cell lines we found viral DNA hybridizing to four fast-moving bands of 5.3, 4.7, 4.4, and 1.3 kb. Such uniformity in the restriction patterns of all cell lines had never been observed

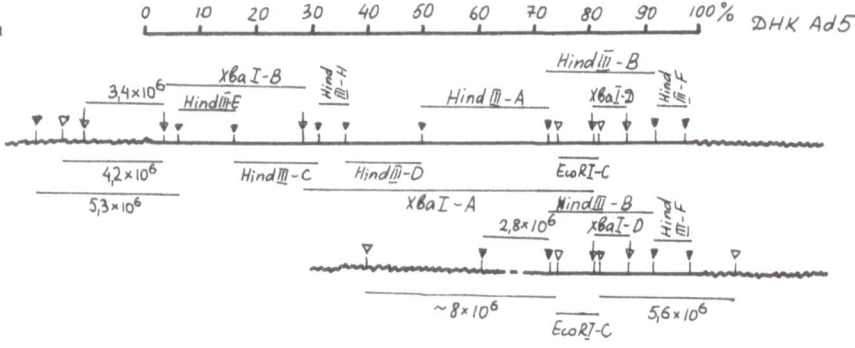

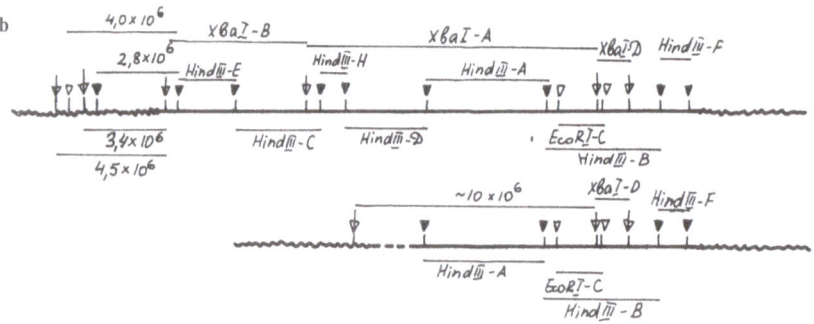

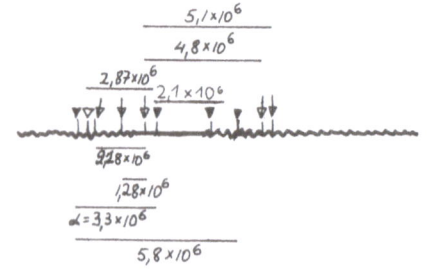

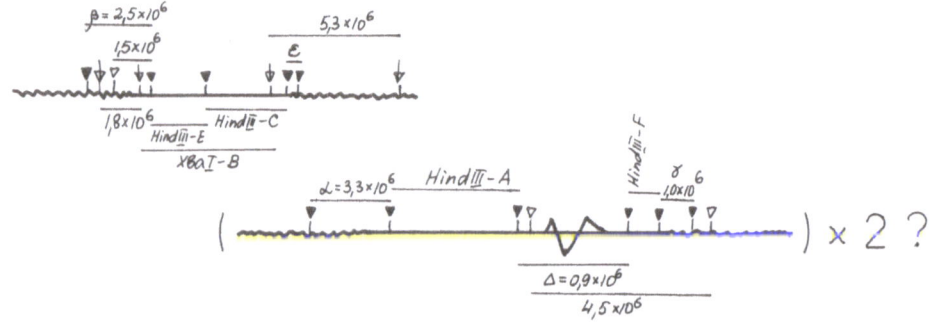

Fig. 3. Summary of the data yielded by analysis of integration patterns in DNA of four cell lines transformed by Ad5 DNA. The most likely models of organization of viral inserts are presented schematically

before when cellular DNA was cut by other restriction enzymes implied that the fast-moving bands might be of cellular origin.

8 Normal Cellular DNA Sequences Homologous to Ad5 DNA

*Bam*HI restriction endonuclease patterns of normal cellular DNA were analyzed to determine the origin of *Bam*HI fast-moving bands unique for different Ad5-transformed cell lines (FROLOVA and GEORGIEV 1979b). It was found that normal rat liver DNA also hybridized to Ad5 DNA and at least three of four fast-moving discrete bands identical with those from Ad5 transformed rat embryo cell hybridized to viral DNA (Fig. 4). Control experiments with cloned Ad5 DNA fragments showed the same results. As shown in Fig. 4, sequences homologous to Ad5 DNA were also found in human DNA, but the organization of such sequences in human DNA differed at least in part from rat DNA.

Nucleic acid homology with the regions of the host genome is not exclusive to Ad5 (JONES and SHENK 1978). It has been demonstrated for Ad2 (JONES

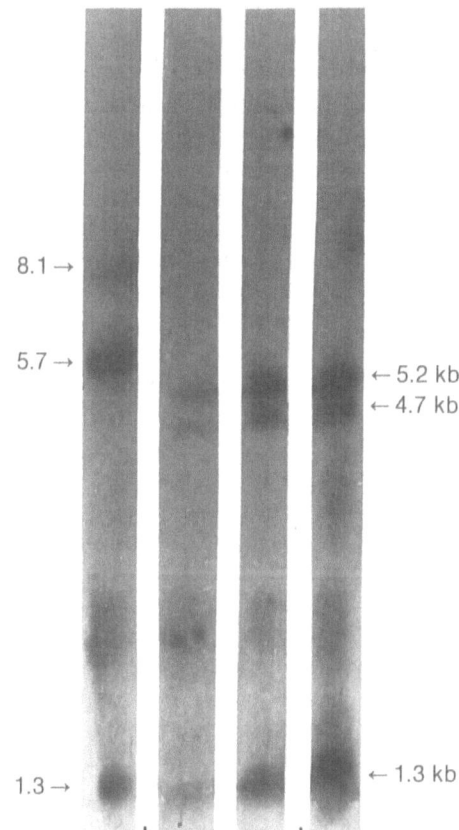

Fig. 4a–d. Hybridization of *Bam*HI fragments of human placenta DNA (**a**) and normal rat liver DNA (**b, c, d**) with ^{32}P-labeled DNA of Ad5 (**a, d**). Hind III-G fragment of Ad5 DNA (**b**) and recombinant plasmid pBR322 containing *Hind*III-E fragment of Ad5 DNA (**c**)

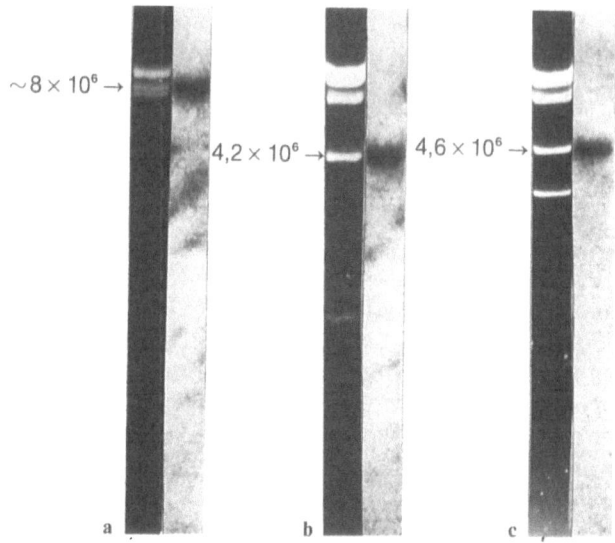

$\sim 8 \times 10^6 \rightarrow$

$4,2 \times 10^6 \rightarrow$ $4,6 \times 10^6 \rightarrow$

a   b   c

Fig. 5a–c. Hybridization of ^{32}P-labeled Ad5 DNA to Southern filters containing EcoRI fragments of clones 5 (**a**); 8 (**b**), and 16 (**c**)

et al. 1979), simian virus 40 (McCutchan and Singer 1981; Queen et al. 1981), herpes simplex virus (Peden et al. 1982; Puga et al. 1982), and the retroviruses (Stehelin et al. 1976; Frenkel et al. 1979; Oskarsson et al. 1980). The biological significance of this phenomenon may be different for each virus-host sequence system and understanding of its significance for retroviruses is in the very early stages.

To isolate normal rat DNA sequences homologous to Ad5 DNA, we screened about 7×10^5 clones of the rat genome library in the Charon 4A phage for hybridization with Ad5 DNA. About 20 strong positives were detected, but after four cycles of recloning only four different clones that contained sequences hybridizing to Ad5 DNA were isolated. One of them binds much less Ad5 DNA than the others and has not been studied, while three clones binding significant amounts of Ad5 DNA were selected for detailed examination. They were designated c15, c18, and cl16.

For the construction of the rat genome library EcoRI underrestricted DNA was used. Therefore characterization of clones started with measurement of the size of EcoRI fragments. The EcoRI fragments of different clones were quite different in size and number. The total size of rat DNA insertions varies from 7.5×10^6 to 20×10^6 daltons and the number of EcoRI restriction fragments in insertion from two to four segments. The EcoRI fragments of the phages were transferred to a nitrocellulose filter and hybridized with ^{32}P-labeled Ad5 DNA (Fig. 5). In each case only one EcoRI fragment of a rather high molecular weight bound the radioactivity.

Experiments of the same type were then performed with clones restricted by BamHI endonuclease (Fig. 6). With this enzyme more restriction fragments were formed. Again, for each clone only one of the fragments hybridized to Ad5 DNA. The size of hybridizing fragments was lower: 5.3, 2.65, and 3.15 kb for cl5, cl8, and cl16 respectively.

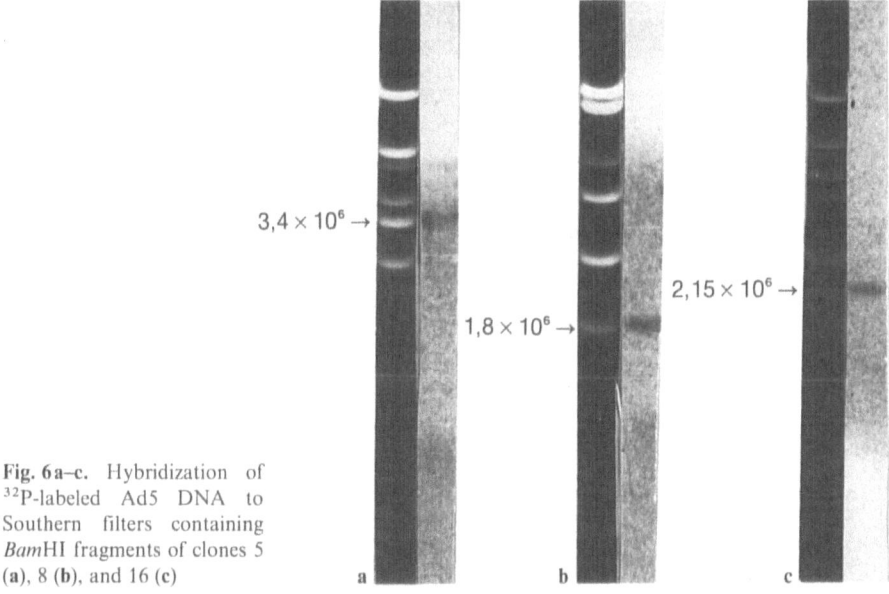

Fig. 6a–c. Hybridization of ^{32}P-labeled Ad5 DNA to Southern filters containing *Bam*HI fragments of clones 5 (a), 8 (b), and 16 (c)

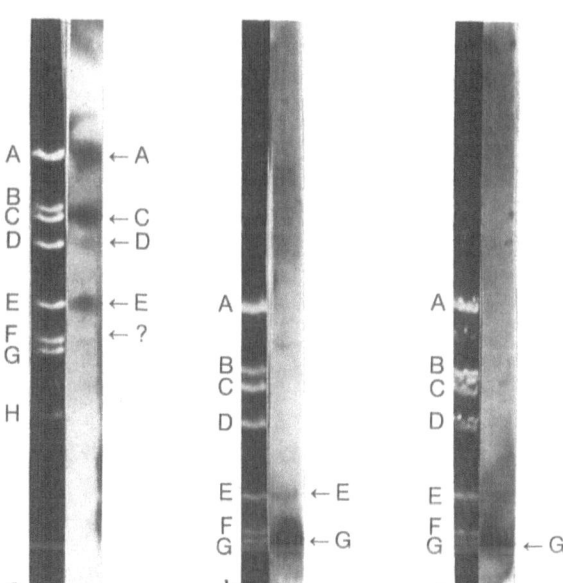

Fig. 7a–c. Hybridization of ^{32}P-labeled DNA from clones 5 (b), 16 (a), and 8 (c) to Southern filters containing *Hin*dIII fragments of Ad5 DNA

To find the regions of the Ad5 genome to which rat DNA sequences do hybridize, Ad5 DNA was restricted with *Hin*dIII endonuclease and the Southern filters containing these fragments were hybridized to the labeled DNA of the clones (Fig. 7). DNA of cl16 hybridizes to a number of *Hin*dIII fragments

```
BspRI                     HhaI              AluI          50
GGCCTTGCACTT  CCTAGGTAAG  CGCTCTACCA  CTGAGCTAAA  TCCCCAGCCC
CCGGAACGTGAA  GGATCCATTC  GCGAGATGGT  GACTCGATTT  AGGGGTCGGG

                                                        100
GTTACTGCAA  CTTTTAAGGA  TTCTTCCATA  GTTTTATGTC  TGGATTGTTT
GAATGACGTT  GAAAATTCCT  AAGAAGGTAT  CAAAATACAG  ACCTAACAAA

                                                        150
CCTTTCTTTC  CTCTCTCTCT  CTCTCTCTCT  CTCTCTCTCT  CTCTCTCTCT
GGAAAGAAAG  GAGAGAGAGA  GAGAGAGAGA  GAGAGAGAGA  GAGAGAGAGA

                                                        200
CATCTTTGTG  TGTGGGAGTT  GGCATAGAAG  CCTTCAATGT  CTCTCATTCT
GTAGAAACAC  ACACCCTCAA  CCGTATCTTC  GGAAGTTACA  GAGAGTAAGA

          HinfI PstI                          HindIII  250
AAGATACGAG  TCTGCAGTCA  CGCATACCAT  TGCAAAACAA  GCTTCCTTGT
TTCTATGCTC  AGACGTCAGT  GCGTATGGTA  ACGTTTTGTT  CGAAGGAACA

        AluI                    Sau3AI                300
CCATAGCAGC  TGGGGCGGCA  CAGATCACGG  GCATCGGCAT  GGTTTAATGG
GGTATCGTCG  ACCCCGCCGT  GTCTAGTGCC  CGTAGCCGTA  CCAAATTACC

                  MspI                                350
GGACACGATA  GACTCCGGAA  ATCTTTTGTG  GAGACTTAAC  CCAGAAAATG
CCTGTGCTAT  CTGAGGCCTT  TAGAAAACAC  CTCTGAATTG  GGTCTTTTAC

                          Sau96I        HinfI        400
AAATTGTCTT  CATCTCGGAG  GAGGACCTGT  GTGGACTCCA  TGAGGCTGCG
TTTAACAGAA  GTAGAGCCTG  CTCCTGGACA  CACCTGAGGT  ACTCCGACGC

HhaI                                          AluI  450
CACCACCTTG  TCCTTGGTGC  TGCTGTTCCA  CCTGAAAGCC  CAGCTCTGTC
GTGGTGGAAC  AGGAACCACG  ACGACAAGGT  GGACTTTCGG  GTCGAGACAG

                                                        500
CTGTTGTCGC  CCTTGGGCAG  AGTTGTGCCT  CCTGCTCTTC  TTTCCCTAAG
GACAACAGCG  GGAACCCGTC  TCAACACGGA  GGACGAGAAG  AAAGGGATTC

                                                        550
GAGGGGCAGC  AGCAGCAGCA  GCAGGAGGAG  CAGGAGGAGC  AGCAGCAGGA
CTCCCCGTCG  TCGTCGTCGT  CGTCCTCCTC  GTCCTCCTCG  TCGTCGTCCT

                                                        600
GCAGCAGGAG  CAGCAGCAGC  AGCAGGAGGA  GCAGGAGCAG  CAGCAGGAGC
CGTCGTCCTC  GTCGTCGTCG  TCGTCCTCCT  CGTCCTCGTC  GTCGTCCTCG

                                                        650
AGCAGCAGCA  GGAGCAGCAG  CAGCAGCAGG  AGCAGGAGGA  GCAGCAG...
TCGTCGTCGT  CCTCGTCGTC  GTCGTCGTCC  TCGTCCTCCT  CGTCGTC

                                                       1080
AGCAGCAGGA  GCAGCAGGAG  GAGCAGCAGG  AGCAGGAGCA  GCAGGAGCAG
TCGTCGTCCT  CGTCGTCCTC  CTCGTCGTCC  TCGTCCTCGT  CGTCCTCGTC

                                                       1130
CAGCAGGAGC  AGGAGCAGGA  GCAGCAGGAG  CAGCAGGAGC  AGCAGCAGGA
GTCGTCCTCG.:TCCTCGTCCT  CGTCGTCCTC  GTCGTCCTCG  TCGTCGTCCT

                                        AluI         1180
GCAGCAGCAG  CAGCAGCAGC  AGCAGCGGTG  CAGCTCCATG  CCATGGGCC
CGTCGTCGTC  GTCGTCGTCG  TCGTCGCCAC  GTCGAGGTAC  GGTACCCGG
                                                    BspRI
```

Fig. 8. Nucleotide sequence analysis of *BspI*-A fragment of c15 containing the region of homology to Ad5 DNA

of Ad5 DNA. Fragments A, C, and E are strongly labeled, while D binds a small amount of radioactivity.

Fragment H, located between C and D, is not labeled. Thus, DNA of cl16 is homologous to at least two different regions of the Ad5 genome located in the E+C fragments and in the A fragment (including a small part of D). The distance between fragments A and E in the Ad5 genome is equal to 11.9 kb, i.e., much greater than the size of the *Bam*HI fragment of cl16 hybridizing to Ad5 DNA (5.2 kb). In other words, the homologous sequences of cl16 and Ad5 are not colinear.

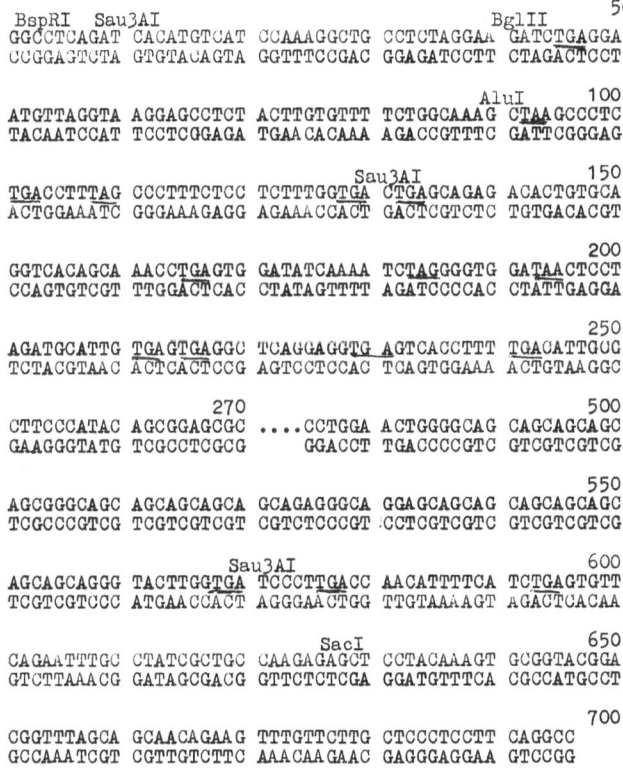

```
                                                                    50
BspRI   Sau3AI                                        BglII
GGCCTCAGAT CACATGTCAT CCAAAGGCTG CCTCTAGGAA GATCTGAGGA
CCGGAGTCTA GTGTACAGTA GGTTTCCGAC GGAGATCCTT CTAGACTCCT

                                                 AluI    100
ATGTTAGGTA AGGAGCCTCT ACTTGTGTTT TCTGGCAAAG CTAAGCCCTC
TACAATCCAT TCCTCGGAGA TGAACACAAA AGACCGTTTC GATTCGGGAG

                                     Sau3AI             150
TGACCTTTAG CCCTTTCTCC TCTTTGGTGA CTGAGCAGAG ACACTGTGCA
ACTGGAAATC GGGAAAGAGG AGAAACCACT GACTCGTCTC TGTGACACGT

                                                       200
GGTCACAGCA AACCTGAGTG GATATCAAAA TCTAGGGGTG GATAACTCCT
CCAGTGTCGT TTGGACTCAC CTATAGTTTT AGATCCCCAC CTATTGAGGA

                                                       250
AGATGCATTG TGAGTGAGGC TCAGGAGGTG AGTCACCTTT TGACATTGCG
TCTACGTAAC ACTCACTCCG AGTCCTCCAC TCAGTGGAAA ACTGTAAGGC

           270                                         500
CTTCCCATAC AGCGGAGCGC ....CCTGGA ACTGGGGCAG CAGCAGCAGC
GAAGGGTATG TCGCCTCGCG      GGACCT TGACCCCGTC GTCGTCGTCG

                                                       550
AGCGGGCAGC AGCAGCAGCA GCAGAGGGCA GGAGCAGCAG CAGCAGCAGC
TCGCCCGTCG TCGTCGTCGT CGTCTCCCGT CCTCGTCGTC GTCGTCGTCG

           Sau3AI                                      600
AGCAGCAGGG TACTTGGTGA TCCCTTGACC AACATTTTCA TCTGAGTGTT
TCGTCGTCCC ATGAACCACT AGGGAACTGG TTGTAAAAGT AGACTCACAA

                         SacI                          650
CAGAATTTGC CTATCGCTGC CAAGAGAGCT CCTACAAAGT GCGGTACGGA
GTCTTAAACG GATAGCGACG GTTCTCTCGA GGATGTTTCA CGCCATGCCT

                                                       700
CGGTTTAGCA GCAACAGAAG TTTGTTCTTG CTCCCTCCTT CAGGCC
GCCAAATCGT CGTTGTCTTC AAACAAGAAC GAGGGAGGAA GTCCGG
```

Fig. 9. Nucleotide sequence analysis of *Bsp*I-A fragment of c18 containing the region of homology to Ad5 DNA

In contrast to cl16, the DNA of cl5 and cl8 hybridizes mostly to only the *Hin*dIII fragment G located at the left end of the adenoviral genome. With the DNA of cl16 some label can also be seen in the adjacent sequences of fragment E.

We determined the primary nucleotide sequence of the regions homologous to Ad5 DNA of the two clones cl5 and cl8 (Figs. 8 and 9). It was found that the *Bsp*I fragment of cl5 hybridizing to Ad5 DNA contains a region approximately 700 bp long and with a simple structure. This simple sequence is composed of repetitive elements CCT and GCT, which are interspersed without any evident regularity. Comparison of this structure with the known structure of the left end of Ad5 DNA has revealed that Ad5 DNA has a homologous triplet sequence localized in region E1B from nucleotide no. 2150 to no. 2205 (Fig. 10). Three short regions of homology, each 10–15 bp long, that are possibly involved in the formation of the DNA hybrids between normal rat and Ad5 DNA were also found. Analysis of nonsense codons in both structures obtained revealed that all the reading frames are stopped with high frequency (Figs. 8 and 9). This observation excludes the possibility that the region of cellular DNA homologous to the adenoviral genome may code the protein related to the product of Ad5 oncogene. Nevertheless, the possibility exists that short

```
HpaII 2140        2150        2160        2170        2180
      CCGGCG ATAATACCGA CGGAGGAGCA GCAGCAGCAG CAGGAGGAAG
      GGCCGC TATTATGGCT GCCTCCTCGT CGTCGTCGTC GTCCTCCTTC


        2190       2200       2210       2220        HpaII
      CCAGGCGGCG GCGGCAGGAG CAGAGCCCAT GGAACCCGAG AGCCGG
      GGTCCGCCGC CGCCGTCCTC GTCTCGGGTA CCTTGGGCTC TCGGCC
```

Fig. 10. The region of the Ad5 genome containing homology to rat cell DNA

regions of homology between viral and cellular DNA may direct specific integration of viral genome.

Consistent with this possibility are the data of DOERFLER et al. (1982), demonstrating short regions of homology at the site of the viral DNA junction. However, we are still not able to obtain direct evidence concerning involvement of the cloned Ad5 homologous sequences in the process of integration.

The sequences showing homology to adenoviral DNA can be compared to the large number of simple sequences that have been identified within or near several protein-coding genes. The sequence AAGAG has been found to constitute the basic 5-bp unit of one component of a *Drosophila melanogaster* satellite DNA (FRY and BRUTLAG 1979). Other simple sequences that have been characterized are constituted by repeats of dinucleotides (TG or CA) in introns of the human globins (SLIGHTOM et al. 1980), human cardiac actin (HAMADA and KAKUNAGA 1982) and within or downstream of some immunoglobulin genes in mouse (KIM et al. 1981; NIOSHIOKA and LEDER 1980) or CT in 9 spaces between histone genes (SUZES et al. 1978), trinucleotides (TCC and TCA, upstream of some mouse V_H genes (COHEN et al. 1982), tetranucleotides (GATA or GACA, in the sex-specific sequences from the snake *Elaphe radiata* (EPPLEN et al. 1982). A polymorphic region of variable length upstream of the human insulin gene is constituted of repeats of a 14-mer element (BELL et al. 1982).

Some authors have argued that simple DNA sequences present in analogous positions in DNA molecules will reanneal much faster than any of the homologous but unique DNA surrounding it and may act like a zipper (COHEN et al. 1982). It seems to us that it is very important for the role of the regions with simple sequences in adenoviral genomes and the significance of their homology with cellular simple sequences for the integration of viral DNA during transformation to be investigated in greater detail.

References

Adelman SF, Howett MK, Rapp P (1981) Tumorigenicity of herpesvirus-transformed cells correlates with production of plasminogen activator. Mol Cell Biol 1:408–417

Adolf KW, Cheng SM, Laemmly UK (1977) Role of nonhistone proteins in metaphase chromosome structure. Cell 12:805–816

Aoi Y, Yokota M (1978) Alterations in surface glycoproteins and level of sialyltransferase activity of human embryo kidney cells infected with oncogenic adenovirus type 12. Tokoho J Exp Med 125:177–183

Bell GI, Selby MF, Rutter WF (1982) The highly polymorphic region near the human insulin gene is composed of simple tandemly repeating sequences. Nature 295:31–35

Bernards R, Houweling A, Hertoghs JJL, Bos JL, van der Eb AJ (1981) Localization of the oncogenic potencial of adenovirus type 12 (Abstr). Paper presented at the Second Imperial Cancer Research Fund DNA Tumor Virus Meeting, Cambridge, England, no. 173

Bernards R, Houweling A, Schrier PJ, Bos JL, van der Eb AJ (1982) Characterization of cells transformed by Ad5/Ad12 hybrid early region plasmid. Virology 120:422–432

Botchan M, Topp W, Sambrook J (1976) The arrangement or simiam virus 40 sequences in the DNA of transformed cells. Cell 9:269–287

Botchan M, Stringer J, Mitchison T, Sambrook J (1980) Integeration and excision of SV40 DNA from the chromosome of transformed cells. Cell 20:143–152

Bourgaux P, Syllia BS, Chartrand P (1982) Excision of polyoma virus DNA from that of a transformed mouse cell: identification of a hybrid molecule with direct and inverted repeat sequences at the virus-cellular joints. Virology 122:84–97

Bramwell ME, Harris H (1978) An abnormal membrane glycoprotein associated with malignancy in a wide range of different tumours. Proc R Soc Lond [Biol] 201:87–106

Bramwell ME, Harris H (1979) Some further information about the normal membrane glycoprotein associated with malignancy. Proc R Soc Lond [Biol] 203:93–99

Brusca JS, Chinnadurai G (1981) Transforming genes among three different oncogenic subgroups of human adenoviruses have similar replicative functions. J Virol 39:300–305

Byrd PJ, Chia W, Rigby PWJ, Gallimore PH (1982) Cloning of DNA fragments from the left end of the adenovirus type 12 genome: transformation by cloned early region I. J Gen Virol 60:279–293

Chen LB, Gallimore PH, McDougall JK (1976) Correlation between tumor induction and the large external transformation sensitive protein on the cell surface. Proc Natl Acad Sci USA 73:3570–3574

Cohen YB, Effron K, Rechavi G, Ben-Neriah J, Zakut R, Gival D (1982) Simple DNA sequences in homologous flanking regions near immunoglobulin V_H genes: a role in gene interaction? Nucleic Acids Res 10:3353–3360

Della Valle G, Fenton RC, Basilico C (1981) Polyoma large T antigen regulates the integration of viral DNA sequences into the genome of transformed cells. Cell 23:347–355

Deuring R, Winterhoff U, Tamanoi F, Stabel S, Doerfler W (1981a) Site of linkage between adenovirus type 12 and cell DNAs in hamster tumor line. CLAC3. Nature 293:81–84

Deuring R, Stabel S, Winterhoff U, Gahlmann R, Vardimon L, Tamanoi F, Renz D, Doerfler W (1981b) Analysis of the sites of junction between adenovirus DNA and cell DNA in transformed and tumor lines (Abstr). Paper presented at the Second Imperial Cancer Research Fund DNA Tumor Virus Meeting, Cambridge, no. 168

Dijkeme R, Dekker BMM, Van der Gelts MJM, Van der Eb AJ (1979) Transformation of primary rat kidney cells by DNA fragments of weecly oncogenic adenoviruses. J Virol 32:943–950

Doerfler W (1982) Uptake, fixation and expression of foreign DNA in mammalian cells: the organization of integrated adenovirus DNA sequences. Curr Top Microbiol Immunol 101:128–188

Doerfler W, Stabel S, Ibelgaufts H, Sutter D, Neumann R, Groneberg J, Scheidtmann KH, Deuring R, Winterhoff U (1979) Selectivity in integration sites of adenoviral DNA. Cold Spring Harbor Symp Quant Biol 44:551–564

Doerfler W, Kuhlmann I, Winterhoff U, Neumann R, Stabel S, Schirm S, Eick D (1982) Integration, methylation, and expression of adenovirus type 12 DNA in transformed and tumor cells. In: Schöne HH, Winnacher EL (eds) Genes and tumor genes. Raven, New York, pp 25–37

Dorsch-Häsler K, Fischer PB, Weinstein B, Ginsberg HS (1980) Patterns of viral DNA integration in cells transformed by wild type or DNA-binding protein mutants of adenovirus type 5 and effect of chemical carcinogens on integration. J Virol 34:305–314

Eick D, Doerfler W (1982) Integrated adenovirus type 12 DNA in the transformed hamster cell line T 637: sequence arrangements at the termini of viral DNA and mode of amplification. J Virol 42:317–321

Epplen FT, McCarrey FR, Suton S, Ohno S (1982) Base sequence of a cloned shake W-chromosome DNA fragment and identification of a mal-specific putative mRNA in the mouse. Proc Natl Acad Sci USA 79:3798–3802

Fanning E, Doerfler W (1976) Intracellular forms of adenovirus DNA 5. Viral DNA sequences in hamster cells abortively infected and transformed with human adenovirus type 12. J Virol 20:373–383

Fanning E, Baske K, Sutter D, Doerfler W (1978) Selectivity in the integration of viral DNA in cells infected and transformed by adenovirus. In: Hofschneider PH, Starlinger P (eds). Integration and excision of DNA molecules. Springer, Berlin Heidelberg New York, pp 81–91

Flint SJ, Sharp PA (1976) Adenovirus transcription. 5. Quantitation of viral RNA sequences in adenovirus 2-infected and transformed cells. J Mol Biol 106:749–771

Flint SJ, Weintraub HM (1977) An altered subunit configuration associated with the actively transcribed DNA of integrated adenovirus genes. Cell 12:783–794

Franke WW (1977) Membrane changes during neoplastic transformation. In: Koprowski H (ed) Neoplastic transformation. Mechanisms and consequences. (Life Sciences Research Reports 7, Berlin, pp 181–195)

Frenkel N, Locker H, Cox B, Roizman B, Rapp F (1976) Herpes simplex virus DNA in transformed cells: sequence complexity in five hamster cell lines and one derived hamster tumor. J Virol 18:885–893

Frenkel A, Gilbert J, Porzig K, Scolnick E, Aaronson S (1979) Nature and distribution of feline sarcoma virus nucleotide sequences. J Virol 30:821–827

Frolova EI, Georgiev GP (1979a) Mapping of the DNA fragments containing viral genes in the DNA of cells transformed by adenovirus type 5. Viruses of cancer and leucosis (in Russian). Ivanovski Institute of Virology, Moscow, pp 23–25

Frolova EI, Georgiev GP (1979b) The existence of DNA sequences homologous to adenovirus 5 DNA in the genome of normal rat cells. Nucleic Acids Res 7:1419–1428

Frolova EI, Zalmanzon ES (1978) Transcription of viral sequences in cells transformed by adenovirus type 5. Virology 89:347–359

Frolova EI, Zalmanzon ES, Georgiev GP (1977) Transcription of the adenovirus type 5 genome in a line of the transformed cells (in Russian). Dokl Akad Nauk SSSR 237:1226–1229

Frolova EI, Zalmanzon ES, Lucanidin EM, Georgiev GP (1978) Studies of the transcription of viral genome in adenovirus 5-transformed cells. Nucleic Acids Res 5:1–11

Fry K, Brutlag D (1979) Detection and resolution of closely related satellite DNA sequences by molecular cloning. J Mol Biol 135:581–593

Gabrielyan ND, Zalmanzon ES, Ivanov SK, Turetskaya RL (1980) Characterization of sialyltransferases in rat embryo cells transformed in vitro by bovine adenovirus type 3 (BAV-3) before and after a passage in the animal. Viruses of cancer and leucosis (in Russian). Ivanovski Institute of Virology, Moscow, pp 30–31

Gabrielyan ND, Zalmanzon ES, Auesova ZhJ, Ivanov SK (1984) Comparative studies of the sialyc transferases in rat embryo cells transformed by the bovine adenovirus type 3 DNA (BA-3) (in Russian). Biochimia 49:261–271

Gahlmann R, Leisten R, Vardinom L, Doerfler W (1982) Patch homologies and the integration of adenovirus DNA in mammalian cells. EMBO J 1:1101–1104

Gallimore PH (1974) Interactions of adenovirus type 2 with rat embryo cells. Permissiveness, transformation and in vitro characteristics of adenovirus transformed rat embryo cells. J Gen Virol 25:263–273

Gallimore PH, Paraskeva C (1980) A study to determine the reasons for differences in the tumorigenicity of rat cell lines transformed by adenovirus 2 and adenovirus 12. Cold Spring Harbor Symp Quant Biol 44:703–713

Gallimore PH, McDougall JK, Bechen L (1979) Malignant behaviour of three adenovirus 2-transformed brain cell lines and their methyl cellulose selected subclones. Int J Cancer 24:477–484

Graevskaya NA, Strizhachenko NM, Karmisheva VJ, Gumina II, Tufanov AV (1972) Cell lines derived from tumours induced in hamsters by bovine adenovirus type 3 (in Russian). Vopr Onkol 18:79–83

Green MR, Mackey JK, Green M (1977) Multiple copies of human adenovirus 12 genomes are integrated in virus-induced hamster tumor. J Virol 22:238–242

Hamada H, Kakunaga T (1982) Potential z-DNA forming sequences are highly dispersed in the human genome. Nature 298:396–398

Harwood LMJ, Gallimore PH (1975) A study of the oncogenicity of adenovirus type 2 transformed rat embryo cells. Int J Cancer 16:498–508

Huebner RJ, Rowe WP, Lane WT (1962) Oncogenic effects in hamster of human adenovirus types 12 and 18. Proc Natl Acad Sci USA 48:2051–2058

Ibelgaufts H, Doerfler W, Scheidtmann KH, Wechsler W (1980) Adenovirus type 12-induced rat tumor cells of neuroepithelial origin: persistence and expression of the viral genome. J Virol 33:423–437

Igarashi K, Sasada R, Kurokawa T, Niiyama Y, Tsukamoto K, Sugino Y (1978) Biochemical studies on bovine adenovirus type 3.4. Transformation by viral DNA and DNA fragments. J Virol 28:219–226

Jochemsen H, Daniels GSG, Hertoghs JJL, Schrier PI, Van der Elsen PJ, Van der Eb AJ (1982) Identification of adenovirus type 12 gene products involved in transformation and oncogenesis. Virology 122:15–28

Jones N, Shenk T (1978) Isolation of deletion and substitution mutants of adenovirus type 5. Cell 13:181–188

Jones KW, Kinross J, Maitland N, Norval M (1979) Normal human tissues contain RNA and antigens related to infectious adenovirus type 5. Nature 272:274–279

Ketner G, Kelly TJ (1976) Integrated simian virus 40 sequences in transformed cell DNA: analysis using restriction endonucleases. Proc Natl Acad Sci USA 73:1102–1106

Kim S, Davis M, Sinn F, Patten P, Hood L (1981) Antibody diversity: somatic hypermutation of rearranged V_H genes. Cell 27:573–581

Kuhlman J, Doerfler W (1982) Shifts in the extent and patterns of DNA methylation upon explanation and subcultivation of adenovirus type 12-induced hamster tumor cells. Virology 118:169–180

Kuhlman J, Achten S, Rudolph R, Doerfler W (1982) Tumor induction by human adenovirus type 12 in hamsters: loss of the viral genome from adenovirus type 12-induced tumor cells in compatible with tumor formation. EMBO J 1:79–86

Lania L, Hayday A, Fried M (1980) Analysis of polyoma virus transformation (Abstr). Presented at the 1980 Tumor Virus Meeting, Cold Spring Harbor, New York, 106

Lee KC, Mak I (1977) Adenovirus type 12 DNA sequences in primary hamster cells tumors. J Virol 24:408–411

Leslie AGW, Arnott S, Chandrasekaran R, Ratliff RL (1980) Polymorphism of DNA double helices. J Mol Biol 143:49–72

Lloyd CW (1975) Sialic acid and the social behaviour of cells. Biol Rev 50:325–350

Maroteaux L, Heilig R, Dupret D, Mandel JL (1983) Repetitive satellite-like sequences are present within or upstream from 3 avian protein-coding genes. Nucleic Acids Res 11:1227–1243

McCutchan T, Singer MF (1981) DNA sequences similar to those around the simian virus 40 origin of replication are present in the monkey genome. Proc Natl Acad Sci USA 78:95–99

Moriuchi T, Yamashita T, Imamura M, Fujinaga K (1982) Altered properties of tumors induced by adenovirus type 12 DNA fragment transformed cells after growth in immunocompetent rats. Int J Cancer 29:101–105

Mougneau E, Birg F, Rassoulzadegan M, Cusin F (1980) Integration sites and sequence arrangement of SV40 DNA in a homogenous series of transformed rat fibroblast lines. Cell 22:917–927

Murphy D, Clayton CE, Righby PWJ (1981) The structure of the integrated viral DNA in SV40-transformed mouse cells. (Abstr) Presented at second Imperial Cancer Research Fund DNA Tumor Virus Meeting, Cambridge, England, no 185

Neer A, Baran N, Manor H (1983) Integration of polyoma virus DNA into chromosomal DNA in transformed rat cells causes deletion of flanking cell sequences. J Gen Virol 64:69–82

Niiyama Y, Sasada R, Igarashi K, Kurokawa T, Sugino Y (1981) Biochemical studies on bovine adenovirus type 3. V. Some properties of mouse cells transformed with viral DNA fragments. Cell Struct Funct 6:121–131

Nioshioka Y, Leder P (1980) Organization and complete sequence of identical embryonic and plasmacytoma k V-region genes. J Biol Chem 255:3691–3694

Oskarsson M, McClements WL, Blair DG, Maizel JV, Vande Woude GF (1980) Properties of a normal mouse cell DNA sequence (Sarc) homologous to the Src sequence of Moloney sarcoma virus. Science 207:1222–1224

Paraskeva C, Gallimore PH (1980) Tumorigenicity and in vitro characteristics of rat liver epithelial cells and their adenovirus-transformed derivative. Int J Cancer 25:631–639

Paulson JR, Laemmly UK (1977) The structure of histone-depleted metaphase chromosomes. Cell 12:817–828

Peden K, Mounts PH, Hayward GS (1982) Homology between mammalian cell DNA sequences and human herpesvirus genomes detected by a hybridization procedure with high-complexity probe. Cell 31:71–80

Puga A, Cantin FM, Notkins AL (1982) Homology between murine and cellular DNA sequences and the terminal repetition of the S component of herpes simplex virus type 1 DNA. Cell 31:81–87

Queen C, Lord ST, McCutchan TF, Singer MF (1981) Three segments from the monkey genome that hybridize to simian virus 40 have common structural elements. Mol Cell Biol 1:1061–1068

Razin SV, Mantieva VL, Georgiev GP (1978) DNA adjacent to attachment points of deoxyribonucleoprotein fibril to chromosomal axial structure is enriched in reitered base sequences. Nucleic Acids Res 5:4734–4750

Ressons A, Bibor-Hardy V, Suh M, Simard R (1979) Analysis of chromosomes, nucleic acids and polypeptides in hamster cells transformed by Herpes Simplex virus type 2. Cancer Res 39:3225–3234

Richardson CL, Baker SR, Morre DJ, Keenan W (1975) Glycosphingolipid synthesis and tumorigenesis. A role for the Golgi apparatus in the origin of specific receptor molecules of the mammalian cell surface. BBA (Biochim Biophys Acta) Libr 417:175–186

Ruley HE, Lania L, Chaudry F, Fried M (1982) Use of a cellular polyadenylation signal by viral transcripts in polyoma virus transformed cells. Nucleic Acids Res 10:4515–4524

Sambrook J, Botchan M, Gallimore P, Ozane B, Pettersson U, Williams J, Sharp P (1974) Viral DNA sequences in cell transformed by Simian virus 40, adenovirus type 2 and adenovirus type 5. Cold Spring Harbor Symp Quant Biol 39:615–632

Sambrook J, Greene R, Stringer J, Mitchison T, Hu S-L, Botchan M (1979) Analysis of the sites of integration of viral DNA sequences in rat cells transformed by adenovirus 2 or SV40. Cold Spring Harbor Symp Quant Biol 44:569–584

Sharp PA, Pettersson U, Sambrook J (1974a) Viral DNA in transformed cells. 1. A study of the sequences of Ad2 DNA in a line of transformed cells using specific fragments of the viral genome. J Mol Biol 86:708–726

Sharp PA, Gallimore PH, Flint SJ (1974b) Mapping of Ad 2 DNA sequences in lytically infected cells and transformed cells. Cold Spring Harbor Symp Quant Biol 39:457–474

Shiroki K, Hauda H, Shimojo H, Yano S, Ojima S, Fujinaga K (1977) Establishment and characterization of rat cell lines transformed by restriction endonuclease fragments of adenovirus 12 DNA. Virology 82:462–471

Shiroki K, Shimojo H, Sawada Y, Uemizu Y, Fujinaga K (1979) Incomplete transformation of rat cells by a small fragment of adenovirus 12 DNA. Virology 95:127–136

Slighton JL, Blechl AF, Smithies O (1980) Human fetal $^G\gamma$- and $^A\gamma$-globin genes: complete nucleotide sequences suggest that DNA can be exchanged between these duplicated genes. Cell 21:627–638

Southern EM (1975) Detection of specific sequences among DNA fragments separated by gel electrophoresis. J Mol Biol 98:503–517

Stabel S, Doerfler W, Friis RR (1980) Integration sites of adenovirus type 12 in transformed hamster cells and hamster tumor cells. J Virol 36:22–40

Stehelin D, Varmur H, Bishop J, Vogt P (1976) DNA related to the transforming gene(s) of avian sarcoma virus is present in normal avian DNA. Nature 260:170–173

Stiles CD, Desmond W, Chuman LM, Sato G, Saier MH (1976) Relationship of cell behaviour in vitro to tumorigenicity in athymic nude mice. Cancer Res 36:3300–3305

Stringer JR (1981) Integrated simian virus 40 DNA: nucleotide sequences at cell-virus recombinant junctions. J Virol 38:671–679

Strizhachenco NM, Graevskaja NA, Karmisheva VJ, Abramova VP, Sjurin VN (1974) About antigenic nonspecificity of the tumors induced by viruses to the ethyologic agent (in Russian). Selkhos Biol 9:906–911

Sures I, Lowry J, Kedes LH (1978) The DNA sequence of sea urchin (S. purpuratus) H2A, H2B and H3 histone coding and spacer regions. Cell 15:1033–1044

Sutter D, Westphal M, Doerfler W (1978) Patterns of integrations of viral DNA sequences in the genome of adenovirus type 12-transformed hamster cells. Cell 14:569–585

Tikchonenko TI, Kalinina TI, Ponomareva TI, Naroditsky BS (1981a) About homologic sequences between the animal cell DNA and adenovirus type 6 DNA. Viruses of Cancer and Leukosis (in Russian). Ivanovski Institute of Virology, Moskow, pp 101–102

Tikchonenko TI, Dreizin RS, Chaplygina NM, Kalinina TI, Gartel AL, Noriditsky BS, Ponomareva TI, Tjunnikoff GI (1981b) Independent integration into the genome of transformed and tumor cells of the DNA fragments with different source. Viruses of Cancer and Leukosis (in Russian). Ivanovski Institute of Virology, Moskow, pp 98–100

Tikchonenko TI, Chaplygina NM, Kalinina TI, Gartel AL, Ponomareva TI, Naroditsky BS, Dreizin RS (1981c) Integration of foreign genome fragments into cells transformed or contransformed with fragmented adenoviral DNA. Gene 15:349–359

Van Beek WP, Smets LA, Emmelot P (1973) Increased sialic acid density in surface glycoproteins of transformed and malignant cells – a general phenomenon? Cancer Res 33:2913–2922

Van Beek WP, Smets LA, Emmelot P (1975) Changed surface glycoproteins as a marker for malignancy in human leukemic cells. Nature 253:457–460

Van der Elsen PJ, Houweling A, de Pater BSC, Van der Veer JL, Van der Eb AJ (1981) The role of the adenovirus early regions E1a and E1b in transformation. (Abstr) Presented at the second Imperial Cancer Research Fund Tumor Virus Meeting on SV40 polyoma and adenoviruses, Cambridge, 128

Van der Putten H, Terwindt E, Berns A, Jaenish R (1979) The integration sites of endogenous and exogenous Moloney murine leukemia virus. Cell 18:109–116

Vardimon L, Doerfler W (1981) Patterns of integration of viral DNA in adenovirus type 2-transformed hamster cells. J Mol Biol 147:227–246

Visser L, Van Mearschalkerweerd MW, Rozin TH, Wassenaar ADC, Reemst AMCB, Sussenbach JS (1979) Viral DNA sequences in adenovirus-transformed cells. Cold Spring Harbor Symp Quant Biol 44:541–550

Wahl GM, Stern M, Stark GR (1979) Efficient transfer of large DNA fragments from agarose gels to diazobenzyloxymethylpaper and rapid hybridization by using dextran sulfate. Proc Natl Acad Sci USA 76:3683–3687

Warren L, Buck CA, Tuszynski GP (1978) Glycopeptide changes and malignant transformation. A possible role for carbohydrate in malignant behaviour. BBA (Biochim Biophys Acta) Libr 516:97–127

Wold WSM, Green M, Mackey JK, Martin JD, Padgett BL, Walker DL (1980) Integration pattern of human JC virus sequences in two clones of a cell line established from a JC virus-induced hamster brain tumor. J Virol 32:1225–1228

Yogeeswaran G, Sebastian H, Stem BS (1979) Cell surface sialylation of glycoproteins and glycosphingolipids in cultured metastatic variant RNA-virus transformed non-producer BALB/C 3T3 cell lines. Int J Cancer 24:193–202

Zalmanzon ES, Frolova EI, Richter B, Mikhailova LN, Turetskaja RL, Savina AA, Bobrova NR (1979) Isolation and characteristic of 7 lines of rat embryo cells, transformed by adenovirus type 5 and their DNA (in Russian). Mol Biol 13:292–308

Zalmanzon ES, Vinkele RA, Grigorieva LV, Turetskaja RL (1981) Obtaining and biological properties of rat embryo cells transformed by bovine adenovirus type 3 (BAV3) DNA (in Russian). Vopr Onkol 27:61–68

Zalmanzon ES, Vinkele RA, Grigorieva LV, Turetskaja RL (1982) A study of rat embryo cells transformed in vitro by the bovine adenovirus type 3 (BAV3) DNA before and after a passage in the host. Virology 123:420–435

Early and Late Proteins of Adenovirus Type 12: Translation Mapping with RNA Isolated from Infected and Transformed Cells

H. Esche, M. Reuther, and K. Schughart

1 Introduction

Human adenoviruses, especially types 2, 5, and 12, have been the subject of extensive studies because of their ability to cause morphological transformation of cultured cells or, in the case of Ad12, to induce tumors in animals. On the other hand, the adenovirus/cell system has clearly established itself as a useful tool for studying gene expression in mammalian cells. Several precise chemical tools for studying the adenovirus genome have become available in recent years; these include methods for separating restriction enzyme fragments of viral DNA and hybridization methods for detecting viral RNAs. The application of recombinant DNA technology has yielded precise information about the genomic and the messenger RNA sequences expressed by different regions of the genome. Most but not all of the proteins encoded by the adenovirus genome have been characterized by structure, but the function of the proteins is still unresolved.

The adenoviruses contain a linear double-stranded DNA molecule of approximately 23×10^6 daltons (35 kilobases) (Green et al. 1967), which codes for around 50 polypeptides. Lytic infection of human cells by adenoviruses e.g., Ad12, proceeds in at least two distinct phases: an early phase, which includes all events preceding the onset of viral DNA replication (about 10 h post infection, Shimojo et al. 1974), and a late phase including events that depend on the onset of DNA synthesis. During the early period only a limited part of the viral genome (early genes) is expressed (Green 1970; Scheidtmann et al.

Institute of Genetics, University of Cologne, D-5000 Cologne

1975). The late phase leads to a drastic change in viral gene expression; almost the entire genome is expressed in functional mRNA. The host cell macromolecular synthesis is progressively inhibited late in adenovirus infection (GINSBERG et al. 1967).

In rodent cells, e.g., hamster cells, Ad12 undergoes an abortive cycle (DOERFLER 1968, 1969; STROHL 1969). These cells are nonpermissive for virus multiplication. Since viral DNA replication cannot be detected in such cells, the block that prevents virus multiplication must be located in a viral or cellular function required early in infection (DOERFLER and LUNDHOLM 1970). Only early mRNAs and proteins are synthesized in this case (RASKA and STROHL 1972; ORTIN and DOERFLER 1975; ORTIN et al. 1976). In baby hamster kidney cells (BHK21) abortively infected with Ad12, approximately 30% of the cell-associated parental viral DNA is linked covalently to cellular DNA (DOERFLER 1975).

During the abortive infection a small fraction of the cells becomes transformed. Stable transformation of cells by adenoviruses is accompanied by integration of viral DNA into the cell genome (DOERFLER et al. 1974; SAMBROOK et al. 1974; DOERFLER 1975, 1977, 1982). Transformation of cells with purified restriction endonuclease fragments of virus DNA (GRAHAM et al. 1974; VAN DER EB et al. 1977; SHIROKI et al. 1977; SEKIKAWA et al. 1978) and studies on the effect of host-range and deletion mutants on the transforming activity of the virus (GRAHAM et al. 1978) have shown that the viral genes that are involved in cellular transformation are located in approximately the left-most 12% of the virus genome. Adenovirus-transformed cells do not produce virions or viral structural proteins; they do, however, express at least some of the early viral genes (HUEBNER et al. 1963; SHARP et al. 1974; FLINT et al. 1975; ORTIN and DOERFLER 1975; LEVINSON and LEVINE 1977; ACHTEN and DOERFLER 1982; ESCHE 1982; ESCHE and SIEGMANN 1982).

Some of the viral functions expressed in the transformed cells might play a role in the initiation and maintenance of the transformed state. Other early viral gene products seem to be essential for transcription (BERK et al. 1979; JONES and SHENK 1979) or DNA replication (VAN DER VLIET et al. 1975) during productive infection. The identification of the viral gene products in adenovirus-infected and -transformed cells and a better characterization of their functions are therefore of particular interest, and may lead to a clear understanding of the biology of these viruses in different types of host cells.

This review of the Ad12 proteins will focus in particular on the identification, structure, and mapping of the Ad12-specific proteins expressed in infected and transformed cells.

2 Methods for Detection and Mapping of Adenovirus Proteins

Several approaches have been used to identify adenovirus proteins. The E2A DNA-binding protein of Ad2 (VAN DER VLIET and LEVINE 1973; JUNG et al. 1978) and of Ad12 (ROSENWIRTH et al. 1975) has been purified from lysates of infected cells. A comparison of proteins prepared from infected and uninfected cells by PAGE has also led to the identification of a number of virus proteins, most of them expressed late in viral infection (for review, see TOOZE 1981). A more specific assignment of the map positions of the adenovirus proteins was obtained by selecting mRNA with genomic DNA restriction fragments, followed by in vitro translation of these mRNAs (ANDERSON et al. 1974; LEWIS et al. 1979; ESCHE et al. 1979). This approach has the advantage of establishing the viral origin of the proteins (by mRNA hybrid selection) and assigning map positions to individual genes on the viral DNA. Finally, serum from animals bearing adenovirus-induced tumors has provided a source of antibodies directed against several adenovirus proteins (RUSSEL et al. 1967; LEVINSON and LEVINE 1977; ACHTEN and DOERFLER 1982). This latter approach, however, is dependent on the quality of the sera obtained. More recently, monoclonal antibodies and antisera directed against synthetic peptides corresponding to specific parts of the viral proteins have been prepared against several Ad2 and Ad12 proteins (SARNOW et al. 1982; CEPKO et al. 1982; ESCHE and BAUSE 1984). Collectively the experimental procedures outlined above have been used to identify and map adenovirus proteins. The functions of some of these proteins have been deduced either by using mutants that alter the function of a gene product (GINSBERG et al. 1974; JONES and SHENK 1979) or by developing functional assays, such as in vitro DNA replication (CHALLBURG and KELLY 1979; STILLMAN 1981) or in vitro transcription (MANLEY et al. 1979).

3 Identification, Structure, and Mapping of Ad12 Proteins

With the methods described in the previous section, 19 virus-specific early and 22 virus-specific late proteins have been detected so far in cells productively infected with Ad12. Their structure and location on the viral genome will now be reviewed.

3.1 Viral Proteins in Productively Infected Cells

3.1.1 Early Viral Proteins

Identification of early Ad12-specific proteins by comparison of ^{35}S-methionine-labeled extracts from infected and mock-infected cells in SDS-PAGE is difficult, because host protein synthesis remains at a high level early after infection. Therefore, the main early Ad12 proteins synthesized in infected cells have been identified by in vitro translation of selected viral mRNAs (ESCHE et al. 1979; JOCHEMSEN et al. 1980, 1982; ESCHE and SIEGMANN 1982).

Viral mRNA synthesized early (0–10 h) after infection is complementary to about 40% of the viral DNA sequences; precisely, to 12–15% of the l-strand and 25–27% of the r-strand of viral DNA (ORTIN et al. 1976). At least six separate regions (E1A, E1B, E2A, E2B, E3, E4), each with its own specific

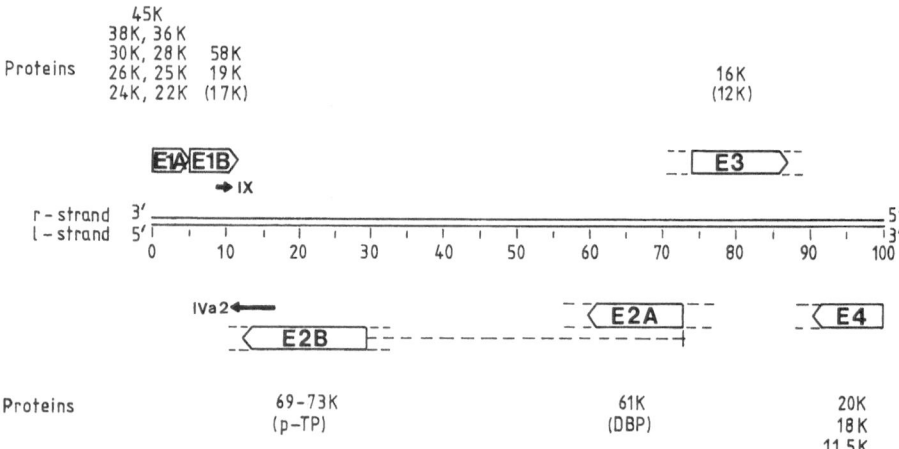

Fig. 1. A genomic map of Ad12-coded proteins detected early in productive infection. Minor protein species are given in *parentheses*. The conventional early gene blocks are shown as *arrows*; the direction of the arrow gives the polarity of transcription in each block. Polypeptides *IX* and *IVa$_2$* (late proteins) can be expressed in the absence of viral DNA replication, but they are most easily detected at late times

promoter or promoters (E1A, E2A) are transcribed at early times after infection (Fig. 1). The map positions of the early regions were determined by hybridization of the mRNAs to various DNA restriction fragments (ORTIN and DOERFLER 1975; SCHEIDTMANN et al. 1975; ORTIN et al. 1976; SAWADA and FUJINAGA 1980). Regions E1 and E3 are transcribed from the r-strand, whereas the coding sequences for regions E2 and E4 are located on the l-strand. Each early region produces a variety of mRNAs, which differ in their splicing pattern. The various transcripts from one region contain overlapping sequences. As a consequence, proteins translated from these mRNAs may also show partial homology. Most early regions of Ad12 DNA indeed specify proteins that are related to one another (ESCHE and SIEGMANN 1982).

Virus-specific mRNA was prepared between 8 and 10 h post infection of human KB cells with Ad12 and was used to direct cell-free protein synthesis. In several experiments cells were treated after infection with inhibitors of DNA (cytosine arabinoside) or protein synthesis (cycloheximide). These treatments prolong the early phase. While inhibitors of DNA synthesis do not enrich viral cytoplasmic RNA efficiently compared with untreated cells, viral RNA synthesis is enhanced about 10- to 20-fold by inhibition of protein synthesis 2–4 h after infection with cycloheximide (CRAIG and RASKAS 1974; EGGERDING and RASKAS 1978). No significant differences were found, however, in the pattern of mRNAs or proteins produced by cell-free translation of mRNAs prepared from cells grown in the presence or absence of these durgs (CHOW et al. 1979; ESCHE and SIEGMANN 1982).

Early proteins synthesized in vitro by RNA selected with different restriction endonuclease fragments of Ad12 DNA and those immunoprecipitated from of extracts of infected cells have been summarized in Table 1 and Fig. 1.

Table 1. Adenovirus type 12 early proteins

Region	Approximate map position	Polarity of transcripts	Proteins
E1A	1.3–4.5	r	45K, 38K, 36K 30K, 28K, 26K, 25K, 24K, 22K
E1B	4.5–11.2	r	58K, 19K (17K)
E2A	61–67	l	61K
E2B	11–30	l	69–73K
E3	76–87	r	16K (12K)
E4	90–99	l	20K, 18K, 11.5K
Late genes expressed to some extent at early times			
IX	9.7–11.2	r	18K
IVa$_2$	11.2–16	l	52K

Messenger RNAs complementary to region E1A (0–4.5 map units) encode at least eight proteins with apparent molecular weights of 38K, 36K, 30K, 28K, 26K, 25K, 24K, and 22K as determined by electrophoresis in SDS-polyacrylamide gels (ESCHE and SIEGMANN 1982; JOCHEMSEN et al. 1980). In addition, a 45K polypeptide encoded by this region has been detected by immunoprecipitation of extracts of infected cells with sera from tumor-bearing animals (JOCHEMSEN et al. 1982) and with monospecific antibodies (ESCHE and BAUSE 1984). The monospecific antibodies were directed against synthetic peptides, which are encoded by sequences located on the 5′ and 3′ ends of the E1A mRNAs. The E1A proteins synthesized in vitro have been characterized by peptide mapping (ESCHE and SIEGMANN 1982). These results indicate that all E1A proteins are structurally related. This observation has been confirmed by the finding that antibodies directed against synthetic peptides are able to immunoprecipitate out from extracts of infected cells almost the same set of proteins that has been shown to be synthesized by cell-free translation of E1A-specific RNA.

At least four mRNA species with partially common sequences have been shown by S1 mapping to be transcribed from region E1A of Ad12 DNA (SAWADA and FUJINAGA 1980). All E1A proteins derived from these RNAs are translated in the same reading frame (SAWADA and FUJINAGA 1980). The way in which the four E1A mRNAs can encode at least eight proteins has not yet been clarified. Perhaps a single primary translation product is produced from each mRNA and some of the proteins are then specifically modified by phosphorylation or proteolytic events. The nucleotide sequence of the E1A region has been determined (SUGISAKI et al. 1980; BOS et al. 1981), and according to these data the E1A proteins have been predicted to be quite rich in proline and glutamic acid residues. All E1A gene products have very acid isoelectric points (ESCHE and SIEGMANN 1982). The molecular weights of the E1A proteins calculated from the nucleotide sequence are all lower than those of the gene products detected by electrophoresis in SDS-polyacrylamide gels. This might be due to the high proline content of the E1A proteins, which can cause alterations in the migration of proteins in this gel system.

Messenger RNA complementary to region E1B of Ad12 (4.5–11.2 map units) directs cell-free synthesis of proteins with apparent molecular weights of 58K, 19K, and a minor species of 17K (JOCHEMSEN et al. 1982; ESCHE and SIEGMANN 1982). Two different, spliced mRNAs are transcribed from region E1B (SAWADA and FUJINAGA 1980; SAITO et al. 1983). The larger 2.2 kb E1B mRNA encodes both the 58K and the 19K proteins, starting at different AUG triplets (BOS et al. 1981). Both proteins are translated in different reading frames and therefore share no common methionine-containing peptides. The shorter 1.0 kb E1B mRNA appears to encode only the 19K polypeptide. It has been suggested that both the E1B 19K and the E1B 58K proteins are synthesized from the 2.2 kb mRNA produced very early after infection, while the E1B 19K protein is synthesized at higher levels from the 1.0 Kb mRNA produced slightly later during infection (BOS et al. 1981). The minor E1B 17K protein is structurally related to the E1B 19K protein (H. ESCHE, unpublished work). Nearly all sera derived from Ad12-induced tumor-bearing animals contain antibodies against the E1B 58K protein, which appears to be the major viral tumor antigen (ACHTEN and DOERFLER 1982; JOCHEMSEN et al. 1982).

In vitro, cytoplasmic viral mRNA complementary to early region E2A (approx. 61–67 map units) programs the synthesis of one major protein of molecular weight 61K (JOCHEMSEN et al. 1980; ESCHE and SIEGMANN 1982). It represents the single-strand specific DNA-binding protein of Ad12 (ROSENWIRTH et al. 1975). RNA initiated at the E2 promoters at approximately 72–75 map units, but selected with an Ad12 DNA fragment carrying sequences from 16.1 to 28 map units, encodes a 69K–73K polypeptide. This protein may represent the precursor of the polypeptide covalently attached to the 5′ end of the Ad12 virion DNA, by analogy with the Ad2 87K protein expressed by region E2B (STILLMAN et al. 1981). An additional E2B protein encoded between approximately 16 and 28 map units, as described for Ad2, an adenovirus-coded DNA polymerase (105K; LICHY et al. 1982), has not yet been detected for Ad12.

For early region E3 (approx. 76–87 map units), one protein with a molecular weight of 16K has been found by in vitro translation of virus-specific mRNA (ESCHE and SIEGMANN 1982). Messenger RNA complementary to region E4 (approx. 90–99 map units) encodes proteins with molecular weights 20K, 18K, and 11.5K (JOCHEMSEN et al. 1980; ESCHE and SIEGMANN 1982). It is not known whether any of the E4 proteins are structurally related.

As of now, limited information is available about the functions of the early viral proteins in the process of virus multiplication. It is thought that proteins specified by region E1 are somehow involved in cellular transformation (GRAHAM et al. 1974; VAN DER EB et al. 1979; VAN DEN ELSEN et al. 1981, to be published). In addition, E1A proteins seem to be involved in regulation of the expression of most of the other early viral genes (BERK et al. 1979; SHENK et al. 1979) and of cellular genes (NEVINS 1982; SCHRIER et al. to be published). The proteins encoded by regions E2A and E2B are required for viral DNA replication, for both initiation and elongation reactions (VAN DER VLIET and

SUSSENBACH 1975; LEVINSON and LEVINE 1977; HORWITZ 1978; STILLMAN et al. 1981). The function(s) of the E3 protein(s) do(es) not appear to be essential to the virus for replication in culture, as deletion mutants that eliminate the entire E3 region of the virus replicate normally in cell culture (JONES and SHENK 1978). However, it has been demonstrated that cytotoxic T lymphocytes can recognize the E3 19K glycoprotein of Ad2 on the surface of infected and transformed cells and reject these cells via T-cell-mediated killing (PERSSON et al. 1979). Thus, the E3 19K glycoprotein may have a function in vivo. As yet, the functions of the proteins encoded by region E4 are unknown.

3.1.2 Late Viral proteins

The late phase in Ad12 infection starts with the onset of viral DNA replication, about 10–12 h after the infection of human KB cells. During the late phase of infection mRNA is transcribed from almost the entire r-strand of the virus DNA (GREEN et al. 1970; ORTIN et al. 1976), whereas only limited l-strand transcription can be observed. Late mRNAs are synthesized in large quantities and can make up as much as 25% of the total mRNA content of an infected cell (PHILIPSON et al. 1975). Almost all late mRNA species found in the cytoplasm are derived from a common large nuclear precursor, which is spliced. The synthesis of the precursor molecule starts at the major late promoter at approximately 16.3 map units (T. BROKER, personal communication) and terminates near the right-hand end of the viral genome (EVANS et al. 1977). After splicing, mRNAs obtained late have a peculiar composition. These mRNAs all share a common tripartite leader that consists of about 200 base pairs derived from viral sequences located at approximately 16.3, 19.6, and 26 map units (T. BROKER, personal communication).

At least five "families" of cytoplasmic mRNA species originating from the major late promoter can be distinguished late in Ad12-infected cells according to their coding sequence (Fig. 2). The Ad12 late mRNAs have not yet been mapped precisely by RNA–DNA heteroduplex/S1 or heteroduplex/electron microscopy methods.

Late viral proteins are made in infected cells in considerable amounts, and are therefore easy to study, especially since the synthesis of host proteins is shut off late after infection (GINSBERG et al. 1967). Most late proteins have been detected by immunoprecipitation with antibodies directed against virion proteins and PAGE (WADELL 1979; ESCHE et al. 1979). Like nonstructural late viral proteins, late proteins have been mapped by in vitro translation of mRNA selected with different restriction endonuclease fragments of Ad12 DNA (REUTHER 1981; REUTHER and ESCHE 1984). The structural proteins II (hexon), III (penton), IV (fiber) and the precursor products of polypeptides VI, VII, and VIII have been identified by tryptic peptide mapping (REUTHER and ESCHE 1984). Two late viral proteins that are found in the virus particle, polypeptides IX and IVa2, are expressed by separate promoters located outside the major late transcription unit (Fig. 2) (PETTERSSON and METHEWS 1977; WILSON et al. 1979; CHOW et al. 1979). Both late promoters are active to some extent at

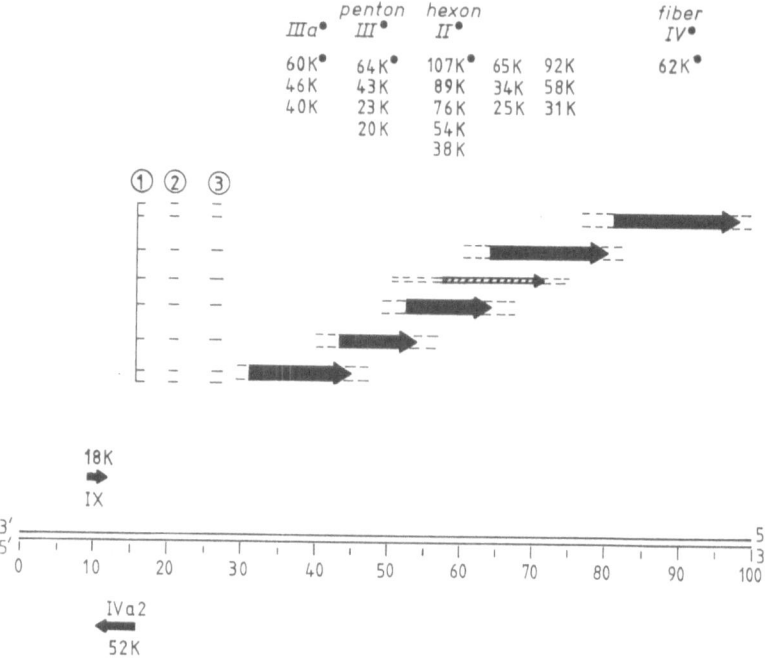

Fig. 2. A genomic map of Ad12-coded proteins detected late in infected cells. The polypeptides have been mapped by cell-free translation of viral mRNAs, which were selected with different restriction endonuclease fragments of Ad12 DNA. *Arrowheads* show the polarity of the transcripts. The relative order of the proteins within each column has not been determined. The *circled numerals* (1–3) mark the approximate loci of the elements of the tripartite leader associated with late mRNA

early times. Transcription from these promoters is greatly enhanced, however, after the onset of viral DNA replication.

3.2 Viral Proteins in Abortively Infected Cells

When human adenoviruses infect human cells the virus multiplies and eventually a large number of progeny particles is released from the cell. Infection of nonhuman cells with human adenoviruses, e.g., rodent cells, usually does not give rise to progeny virus. Some cell types, however, are "semipermissive" for a certain type of virus. In cultures of such cells the virus can multiply only in a small number of cells. Rat cells, for example, are semipermissive for Ad2 and Ad5 multiplication. More frequently, viral DNA cannot be replicated and late proteins are not synthesized either; the infection is then called "abortive." The best studied system is that of BHK21 cells infected with Ad12 (DOERFLER 1968, 1969, 1970; STROHL 1969; DOERFLER et al. 1972; RASKA and STROHL 1972; WEBER and MAK 1972; ORTIN and DOERFLER 1975). Since viral DNA replication cannot be detected in these cells, the block preventing virus multiplication must be located in a viral or cellular function required early in infection.

Ad12 mRNAs that can be isolated from abortively infected BHK21 cells are derived from the same or similar regions of the Ad12 gemome as the mRNAs made early after productive infection (RASKA and STROHL 1972; ORTIN and DOERFLER 1975). No mRNA from late regions can be detected. The proteins directed by cell-free protein synthesis with mRNA prepared from Ad12-infected BHK21 cells are very similar in size to those programmed by mRNA isolated from productively infected cells (ESCHE et al. 1979; H. ESCHE, unpublished data). In BHK21 cells infected with AD12 at high multiplicities, 100% of the cells became positive for T antigen (STROHL 1969). A more detailed analysis of the viral gene products in these cells is required for a better understanding of the basis of abortive infection, however.

3.3 Viral Proteins in Ad12-Transformed Cells

Stable transformation of cells by adenoviruses is accompanied by the integration of viral DNA into the cellular genome. The amount of viral DNA integrated, however, varies from cell line to cell line. A striking observation is that Ad12-transformed cells contain 1–30 copies of sequences from practically the entire viral genome, whereas cells transformed by Ad2 and Ad5 generally contain less then 10 copies of only parts of the virus genome (GALLIMORE et al. 1974; FANNING and DOERFLER 1976; SUTTER et al. 1978; STABEL et al. 1980; DOERFLER 1982). The left 11–14% of the viral genome is integrated into the genome of all fully transformed cells, with a few exceptions. These exceptions are revertants from cultured Ad12-induced tumor cells that have lost all Ad12 DNA sequences (KUHLMANN et al. 1982). Simultaneously with the loss of the Ad12 DNA, the cell morphology changed from epithelioid to fibroblastic. These cells have, however, retained the capacity to induce tumors in animals. Adenovirus-transformed cells with a fully transformed phenotype not only contain but also express viral sequences that cover at least the left-hand end 12% (regions E1A and E1B) of the viral genome (GALLIMORE et al. 1974; GRAHAM et al. 1974; VAN DER EB et al. 1979; ESCHE and SIEGMANN 1982; ACHTEN and DOERFLER 1982). These observations led to the assumption that viral gene products encoded by this region may be responsible for the initiation and maintenance of the transformed phenotype. Additional evidence for this assumption came from the isolation of transformation-defective mutants of human adenoviruses that map within the left-hand end 12% of the viral genome (GRAHAM et al. 1978; JONES and SHENK 1979). There is now general agreement that viral proteins encoded by region E1 are required for cellular transformation.

E1-specific mRNA preparations isolated from different Ad12-transformed rodent cell lines (hamster, rat, and mouse cell lines) direct the synthesis of proteins very similar if not identical in size to those obtained when viral mRNA prepared early from infected cells was translated in vitro (JOCHEMSEN et al. 1982; ESCHE and SIEGMANN 1982; STARZINSKI-POWITZ et al. 1982). An additional protein of 34K, however, has been found only with mRNA prepared from Ad12-transformed cells (Fig. 3) (ESCHE and SIEGMANN 1982). This 34K protein shares common tryptic peptides with E1A proteins and can be precipitated from ex-

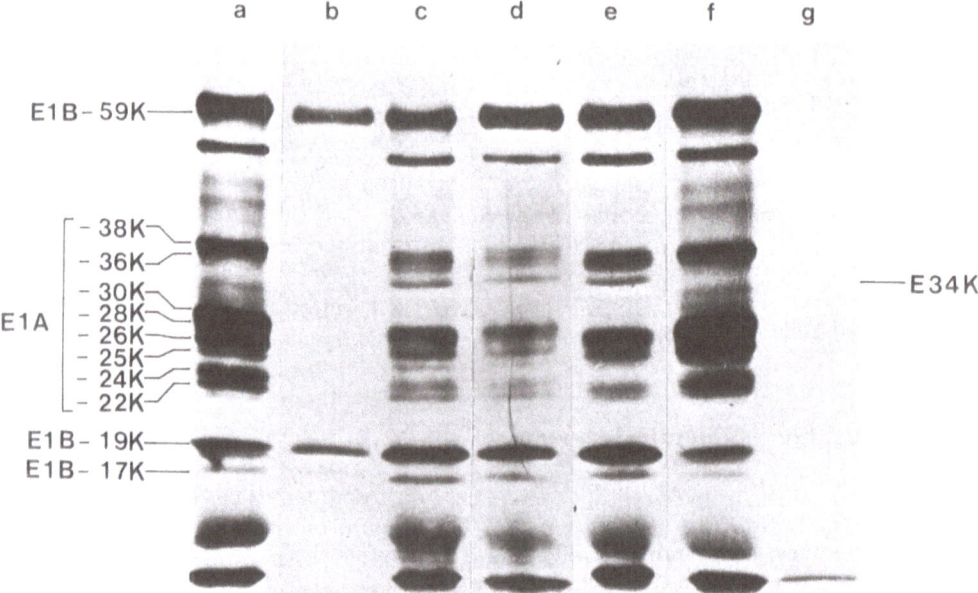

Fig. 3. Early region E1-coded proteins synthesized in vitro by cell-free extracts of rabbit reticulocytes programmed with E1-specific Ad12 RNA prepared from Ad12-transformed hamster and rat cell lines. Messenger RNA was selected from the Ad12-transformed established hamster cell line T637 (*c*), the Ad12-transformed primary Syrian hamster cell line HA12/7 (*d*), and the Ad12-induced rat brain tumor line RBT12/3 (*e*). The products of cell-free translation of the selected RNAs were analyzed by electrophoresis through 12% polyacrylamide gels followed by fluorography. Results of in vitro translation with no added RNA (*g*) and with mRNA prepared from infected cells by hybridization to restriction fragments of the Ad12 DNA-carrying regions E1 (*a, f*) and E1B (*b*) are shown for comparison

tracts of transformed cells with antibodies directed against synthetic peptides which correspond to amino acid sequences located at the C-terminal part of the Ad12 E1A proteins (ESCHE and BAUSE 1984). One possible explanation for the synthesis of the E1A 34K polypeptide in transformed cells might be that the corresponding mRNA species is spliced differently in transformed cells than the RNAs in infected cells. It has been shown that the processing of primary transcripts to cytoplasmic mRNA can follow several alternative pathways with different intermediates (CHOW et al. 1979). Such small variations in virus-specific RNA populations in the transformed cells may be caused by differences in the growth properties of these cells.

In addition to the left-hand 12% (region E1), other regions of the viral genome, which are not required for transformation, are also integrated into adenovirus-transformed cell lines (SUTTER et al. 1978; DOERFLER et al. 1979; SAMBROOK et al. 1979; DOERFLER 1982). It has been reported by several groups (FLINT et al. 1976; LEVINSON and LEVIN 1977; ORTIN et al. 1976; ESCHE 1982; ESCHE and SIEGMANN 1982) that viral RNA and proteins encoded by regions outside the "transforming region" were indeed found in transformed cells. Our results and those of other investigators suggest that the conventional early regions E1, E2A, E3, and E4 are expressed in Ad12- or Ad2-transformed cells whenever they are present. This is not unexpected, since transcription of the early regions does not require preceding replication of viral DNA. As the length of each of the early transcription units corresponds to less then 10 map units (with the exception of early region E2B), the chance that these small transcription units are integrated intact is rather high, even in cell lines in which the viral DNA is not integrated intact but in relatively short fragments.

Proteins expressed by early regions E2A, E3, and E4 of Ad12-transformed cells whenever the early regions are present are similar in size to those found in infected cells (JOCHEMSEN et al. 1980; ESCHE and SIEGMANN 1982). In none of the Ad12-transformed cell lines investigated so far have virus mRNAs and proteins been detected that were expressed by DNA sequences between approximately 11 and 28 map units (region E2B). Neither functional mRNA coding for proteins IX and IVa_2 nor RNA from other late viral regions has been detected in any of the Ad12-transformed cell lines (SCHIRM and DOERFLER 1981; ESCHE and SIEGMANN 1982).

In summary, the general situation at present is that at least in Ad12-transformed cells early regions are expressed, provided they carry a functional promoter. Analogous observations have been made for Ad2-transformed cells (FLINT et al. 1975; ESCHE 1982). However, a few Ad12-transformed cell lines, e.g., cell line HA12/7, do not express all early regions although they are present. This finding might be explained by small deletions or base-exchange mutants that have not yet been detected within the promoter region. One reason why late promoters such as those used for transcription of component IX, IVa_2, the E2 promoter used late after infection, and the major late promoter are inactive in transformed cells might be that these promoter sequences are methylated (DOERFLER 1981; DOERFLER 1983; VARDIMON et al. 1983).

4 Conclusions

To date, about 19 early and 22 late viral proteins have been detected in cells productively infected with Ad12 (Figs. 1 and 2; Table 1). The coding sequences for these proteins have been mapped on the viral genome by cell-free translation of viral mRNAs, which have been selected by hybridization to different restriction endonuclease fragments of Ad12 DNA. It is most likely, however, that not all Ad12-coded proteins have yet been identified. The organization of the viral genes on the Ad12 genome closely resembles that of viral genes on the Ad2 and Ad5 genomes. Less is known about the biochemical structure at least of the early Ad12 proteins. It has been shown that some of these proteins are phosphorylated; the extent of phosphorylation and the location of the phosphor groups, however, are still unknown. Although extensive data have been collected in recent years on the proteins encoded by regions E1A, E1B, E2A, and E2B (for review, see LEVINE 1984), little is known about the function(s) of most early proteins. How do early proteins act in transcription and viral

DNA replication? How do viral proteins interact with cellular proteins and functions? Another important question, namely the question as to how viral gene products of region E1 are involved in oncogenic cellular transformation, has obviously not yet been solved. Therefore the adenovirus/cell system will surely continue to be one of the attractive fields in molecular biology research.

Acknowledgments. We wish to express our thanks to Dr. WALTER DOERFLER for a critical review of the manuscript and to numerous colleagues for providing preprints of unpublished work. The expert and patient secretarial assistance of BIRGIT KIRSPEL and PETRA BÖHM is gratefully acknowledged. Work in our laboratory has been supported by the Deutsche Forschungsgemeinschaft through grant no. SFB74.

References

Achten S, Doerfler W (1982) Virus-specific proteins in adenovirus type 12-transformed and tumor cells as detected by immunoprecipitation. J Gen Virol 59:357–366

Anderson CW, Lewis JB, Atkins JF, Gesteland RF (1974) Cell-free synthesis of adenovirus 2 proteins programmed by fractioned messenger RNA: a comparison of polypeptide products and messenger RNA lengths. Proc Natl Acad Sci USA 71:2756–2760

Berget AM, Sharp PA (1979) Structure of late adenovirus 2 heterogenous nuclear RNA. J Mol Biol 129:547–565

Berk AJ, Lee F, Harrison T, Williams J, Sharp PA (1979) Preearly adenovirus 5 gene product regulates synthesis of early viral messenger RNAs. Cell 17:1935–1944

Bos JL, Polder LJ, Bernards R, Schrier PI, van der Elsen PJ, van der Eb AJ, van Ormondt H (1981) The 2.2 kb Elb mRNA of human Ad12 and Ad5 codes for two tumor antigens starting at different AUG triplets. Cell 27:121–131

Cepko CL, Chanelian PS, Sharp PH (1982) Immunoprecipitation with two-dimensional pools as a hybridoma screening technique: production and characterization of monoclonal antibodies against adenovirus 2 proteins. Virology 11:385–401

Challberg M, Kelly TS (1979) Adenovirus DNA replication in vitro. Proc Natl Acad Sci USA 76:655–659

Chow LT, Roberts JM, Lewis JB, Broker TR (1977) A map of cytoplasmic RNA transcripts from lytic adenovirus type 2, determined by electron microscopy of RNA:DNA hybrids. Cell 11:819–836

Chow LT, Broker TR, Lewis JB (1979) Complex splicing patterns of RNAs from the early regions of adenovirus 2. J Mol Biol 14:265–304

Craig EA, Raskas HJ (1974) Effect of cycloheximide on RNA metabolism early in productive infection with adenovirus 2. J Virol 14:26–32

Doerfler W (1968) The fate of DNA of adenovirus type 12 in baby hamster kidney cells. Proc Natl Acad Sci USA 60:636–643

Doerfler W (1969) Nonproductive infection of baby hamster kidney cells (BHK21) with adenovirus type 12. Virology 38:587–606

Doerfler W (1970) Integration of DNA of adenovirus type 12 into the DNA of baby hamster kidney cells. J Virol 6:652–666

Doerfler W (1975) Integration of viral DNA into the host genome. In: Current Topics in Microbiology and Immunology Vol 71. Springer, Berlin Heidelberg New York, pp 1–78

Doerfler W (1977) Animal virus-host genome interactions. In: Fraenkel-Conrat H, Wagner RR (eds) Comprehensive virology. Plenum, New York, vol 10, pp 279–399

Doerfler W (1981) DNA methylation – a regulatory signal in eucaryotic gene expression. J Gen Virol 57:1–20

Doerfler W (1982) Uptake, fixation and expression of foreign DNA in mammalian cells: the organization of integrated adenovirus DNA sequences. In: Graf T, Jaenisch R (eds) Tumorviruses, neoplastic transformation and differentiation. Springer, Berlin Heidelberg New York, pp 127–194 (Current topics in microbiology and immunology, vol 101)

Doerfler W (1983) DNA methylation and gene activity. Annu Rev Biochem 52:1–20

Doerfler W, Lundholm U (1970) Absence of replication of the DNA of adenovirus type 12 cells. Virology 40:754–757

Doerfler W, Lundholm U, Hirsch-Kaufmann M (1972) Intracellular forms of adenovirus deoxyribonucleic acid. I. Evidence for a deoxyribonucleic acid-protein complex in baby hamster kidney cells infected with adenovirus type 12. J Virol 9:297–308

Doerfler W, Burger H, Ortin J, Fanning E, Brown DT, Westphal M, Winterhoff U, Weiser B, Schick J (1974) Integration of adenovirus DNA into the cellular genome. Cold Spring Harbor Symp Quant Biol 39:505–521

Doerfler W, Stabel S, Ibelgaufts H, Sutter D, Neumann R, Groneberg J, Scheidtmann KH, Deuring R, Winterhoff U (1979) Selectivity in integration sites of adenoviral DNA. Cold Spring Harbor Symp Quant Biol 44:551–564

Eggerding F, Raskas HJ (1978) Effect of protein synthesis inhibitors on viral RNAs synthesized early in adenovirus type 2 infection. J Virol 25:453–458

Esche H (1982) Viral gene products in adenovirus type 2-transformed hamster cells. J Virol 41:1076–1082

Esche H, Bause E (1984) Antibodies specific for amino- and carboxy-terminal regions of adenovirus type 12 tumor antigens. Virus Research (submitted)

Esche H, Siegmann B (1982) Expression of early viral gene products in adenovirus type 12-infected and -transformed cells. J Gen Virol 60:99–113

Esche H, Schilling R, Doerfler W (1979) In vitro translation of adenovirus type 12-specific mRNA isolated and transformed cells. J Virol 30:21–31

Evans RM, Fraser N, Ziff E, Weber J, Wilson M, Darnell JE (1977) The initiator sites for RNA transcription in Ad2 DNA. Cell 12:733–739

Fanning E, Doerfler W (1976) Intracellular forms of adenovirus DNA. V. Viral DNA sequences in hamster cells abortively infected and transformed with human adenovirus type 12. J Virol 20:373–383

Flint SJ, Gallimore PH, Sharp PH (1975) Comparison of viral RNA sequences in adenovirus-transformed and lytically infected cells. J Mol Biol 96:47–68

Flint SJ, Sambrook J, Williams JF, Sharp PH (1976) Viral nucleic acid sequence in transformed cells. IV. A study of the sequences of adenovirus 5 DNA and RNA in four lines of adenovirus 5-transformed rodent cells using specific fragments of the viral genome. Virology 72:456–470

Gallimore PH, Sharp PA, Sambrook J (1974) Viral DNA in transformed cells. II. A study of the sequences of adenovirus 2 DNA in nine lines of transformed rat cells using specific fragments of the viral genome. J Mol Biol 89:49–72

Ginsberg HS, Bello LJ, Levine AJ (1967) Control of biosynthesis of host macromolecules in cells infected with adenoviruses. In: Colter JS, Paranchych W (eds) The molecular biology of adenoviruses. Academic, New York, pp 547–573

Ginsberg HS, Ensinger MS, Kauffman RS, Mayer AJ, Londholm U (1974) Cell transformation: a study of regulation with types 5 and 12 adenovirus temperature-sensitive mutants. Cold Spring Harbor Symp Quant Biol 39:412–426

Graham FL, Abrahams PJ, Mulder C, Heijneker HL, Warnaar SO, de Vries FAJ, Fiers W, van der Eb AJ (1974) Studies on in vitro transformation by DNA fragments of human adenoviruses and simian virus 40. Cold Spring Harbor Symp Quant Biol 39:637–650

Graham FL, Harrison T, Williams J (1978) Defective transforming capacity of adenovirus type 5 host-range mutants. Virology 86:10–21

Green M (1970) Oncogenic viruses. Annu Rev Biochem 39:701–756

Green M, Parson TJ, Pina M, Fujinaga K, Caffier H, Landgraf-Leurs IM (1970) Transcription of adenovirus genes in productively infected and in transformed cells. Cold Spring Harbor Symp Quant Biol 35:803–818

Green M, Pina M, Kimes R, Wensink PC, Mac Hattie LA, Thomas CA (1967) Adenovirus DNA I. Molecular weight and conformation. Proc Natl Acad Sci USA 57:1302–1311

Horwitz MS (1978) Temperature sensitive replication of H5ts125 adenovirus DNA in vitro. Proc Natl Acad Sci USA 75:4291–4295

Huebner RJ, Rowe WP, Turner HC, Lane WT (1963) Specific adenovirus complement-fixing antigens in virus-free hamster and rat tumors. Proc Natl Acad Sci USA 50:379–389

Jochemsen H, Daniels GSG, Kupker H, van der Eb AJ (1980) Identification and mapping of early gene products of adenovirus type 12. Virology 105:172–188

Jochemsen H, Daniels GSG, Hertoghs JJL, Schrier PI, van den Elson PJ, van der Eb AJ (1982) Identification of adenovirus type 12 gene products involved in transformation and oncogenesis. Virology 122:15–28

Jones N, Shenk T (1978) Isolation of deletion and substitution mutants of adenovirus type 5. Cell 13:181–197

Jones N, Shenk T (1979) An adenovirus type 5 early gene function regulates expression of other early viral genes. Proc Natl Acad Sci USA 76:3665–3669

Jung YH, Wold WSM, Sugawara K, Green M (1978) Evidence for an adenovirus type 2 coded early glycoprotein. J Virol 28:314–323

Kuhlmann I, Achten S, Rudolph R, Doerfler W (1982) Tumor induction by human adenovirus type 12 in hamsters: loss of the viral genome from adenovirus type 12-induced tumor cells is compatible with tumor formation. EMBO J 1:79–86

Levine AJ (1984) The adenovirus early proteins. In: Doerfler W (ed) The molecular biology of adenoviruses I. Springer, Berlin Heidelberg New York (Current topics in microbiology and immunology, vol 109)

Levinson AD, Levine AL (1977) The group C adenovirus tumor antigens: identification in infected and transformed cells and a peptide map analysis. Cell 1:871–879

Lewis JB, Esche H, Smart JE, Stillman B, Harter ML, Mathews MB (1979) Organization and expression of the left third of the genome of adenovirus. Cold Spring Harbor Symp Quant Biol 44:493–508

Lichy JH, Field J, Horwitz M (1982) Seperation of the adenovirus terminal protein precursor from its associated DNA polymerase: role of both proteins in the initiation of adenovirus DNA replication. Proc Natl Acad Sci USA 79:5225–5229

Manley JL, Sharp PA, Gefter M (1979) RNA synthesis in isolated nuclei: in vitro mutation of adenovirus 2 major late mRNA precursor. Proc Natl Acad Sci USA 76:160–164

Nevins J (1982) Induction of the synthesis of a 70,000 dalton mammalian heat shock protein by the adenovirus E1A gene product. Cell 29:913–919

Nevins J, Darnall JE (1978) Groups of adenovirus type 2 mRNAs derived from large primary transcript: probably nuclear origin and possible common 3′-ends. J Virol 25:811–823

Ortin J, Doerfler W (1975) Transcription of the genome of adenovirus type 12. I. Viral mRNA in abortively infected and transformed cells. J Virol 15:27–35

Ortin J, Scheidtmann KH, Greenberg R, Westphal M, Doerfler W (1976) Transcription of the genome of adenovirus type 12. III. Maps of stable RNA from productively infected human cells and abortively infected and transformed hamster cells. J Virol 20:355–372

Persson H, Kvist S, Ostberg L, Petersson PA, Philipson L (1979) Adenoviral early glycoprotein E3-191X and its association with transplantation antigens. Cold Spring Harbor Symp Quant Biol 44:509–517

Pettersson U, Mathews MB (1977) The gene and messenger RNA for adenovirus polypeptide IX. Cell 12:741–750

Philipson L, Pettersson U, Lindberg U (1975) Molecular biology of adenoviruses. Virology 14:1–115

Raska K, Strohl WA (1972) The response of BHK21 cells to infection with type 12 adenovirus. VI. Synthesis of Virus-specific RNA. Virology 47:734–742

Reuther-Brötz M (1981) Identifikation und Kartierung der späten Proteine von Adenovirus Typ 12. Thesis, University of Cologne, Cologne, Germany

Reuther M, Esche H (1984, to be published) The map of adenovirus type 12 late proteins. Virology

Rosenwirth B, Shiroki K, Levine AJ, Shimojo H (1975) Isolation and characterization of adenovirus type 12 DNA binding proteins. Virology 67:14–23

Russel WC, Hayashi K, Sanderso PJ, Pereira HG (1967) Adenovirus antigens: a study of their properties and sequential development in infection. J Virol 1:495–507

Saito I, Shiroki K, Shimojo H (1983) Messenger RNA species and proteins of adenovirus type 12 transforming regions: identification of proteins translated from multiple coding stretches in 2.2 kb region 1B mRNA in vitro and in vivo. Virology 127:271–289

Sambrook J, Botchan M, Gallimore P, Ozanne B, Pettersson U, Sharp WJ (1974) Viral DNA sequences in cells transformed by simian virus 40, adenovirus type 2 and adenovirus type 5. Cold Spring Harbor Symp Quant Biol 39:615–632

Sarnow P, Sullivan CA, Levine AJ (1982) A monoclonal antibody detecting the adenovirus 5 Elb-58kd tumor antigen: characterization of the Elb-58kd tumor antigen in adenovirus infected and transformed cells. Virology 34:650–657

Sawada Y, Fujinaga K (1980) Mapping of adenovirus 12 mRNAs transcribed from the transforming region. J Virol 36:639–651

Scheidtmann KH, Ortin J, Doerfler W (1975) Transcription of the genome of adenovirus type 12. II. Viral mRNA in productively infected KB cells. Eur J Biochem 58:283–290

Schirm S, Doerfler W (1981) Expression of viral DNA in adenovirus type 12-transformed cells, in tumor cells, and in revertants. J Virol 39:694–702

Schrier PI, Bernards R, Vaessen RTMJ, Houweling A, van der Eb AJ (1984, to be published) Highly oncogenic adenovirus 12 switches of the expression of class I major histocompility antigens in transformed rat cells. J Immunol

Sekikawa K, Shiroki K, Shimojo H, Ojima S, Fujinaga K (1978) Transformation of a rat cell line by an adenovirus 7 DNA fragment. Virology 88:1–7

Sharp PA, Gallimore PH, Flint SJ (1974) Mapping of adenovirus 2 RNA sequences in lytically infected cells and transformed cell lines. Cold Spring Harbor Symp Quant Biol 39:457–474

Shenk T, Jones N, Colby W, Fowlkes D (1979) Functional analysis of Ad5 host range deletion mutants defective for transformation of rat embryo cells. Cold Spring Harbor Symp Quant Biol 44:367–375

Shimojo H, Shiroki K, Yamaguchi K (1974) The viral DNA replication machinery of adenovirus 12. Cold Spring Harbor Symp Quant Biol 39:533–538

Shiroki K, Handa H, Shimojo H, Yano H, Ojima S, Fujinaga K (1977) Establishment and characterization of rat cell lines transformed by restriction endonuclease fragments of adenovirus 12 DNA. Virology 82:462–471

Stabel S, Doerfler W, Friis RR (1980) Integration sites of adenovirus type 12 DNA in transformed hamster cells and hamster tumor cells. J Virol 36:22–40

Starzinski-Powitz A, Schulz M, Esche H, Mukai N, Doerfler W (1982) The adenovirus type 12 – mouse cell system: permissivity and analysis of integration patterns of viral DNA in tumor cells. EMBO J 1:493–497

Stillman B (1981) Adenovirus DNA-replication in vitro: A protein linked to the 5′ end of nascent DNA strands. J Virol 37:139–147

Stillman B, Lewis JB, Chow LT, Mathews MB, Smart JE (1981) Identification of the gene and mRNA for the adenovirus terminal protein precursor. Cell 23:497–508

Strohl WA (1969) The response of BHK21 cells to infection with type 12 adenovirus. I. Cell killing and T antigen synthesis as correlated viral genome functions. Virology 39:642–652

Sugisaki H, Sugimoto K, Takanami M, Shiroki K, Saito I, Shimojo H, Sawada Y, Uemizu Y, Uesugi SI, Fujinaga K (1980) Structure and gene organization in the transforming HindIII-G fragment of Ad12. Cell 20:777–786

Sutter D, Westphal M, Doerfler W (1978) Pattern of integration of viral DNA sequences in the genomes of adenovirus type 12-transformed hamster cells. Cell 14:569–585

Tooze J (ed) (1981) Molecular biology of tumor viruses. 2nd ed, part 2. DNA tumor viruses. Cold Spring Harbor, New York

van den Elsen PJ, de Pater S, Houwelinh A, van der Veer J, van der Eb AJ (1981) The relationship between region E1A and E1B of human adenoviruses in cell transformation. Gene 18:175–185

van den Elsen PJ, Houweling A, van der Eb AJ (1983) Expression of region E1B of human adenoviruses in the absence of a functional E1A region is not sufficient for complete transformation. Virology

van der Eb AJ, Mulder C, Graham FL, Houweling A (1977) Transformation with specific fragments with transforming activity of adenovirus 2 and 5 DNA. Gene 2:115–132

van der Eb AJ, van Ormondt H, Schrier PI, Lupker JH, Jochemsen H, van den Elsen PJ, De Leys RJ, Kratt J, van Beveren CP, Dykema R, De Wood A (1979) Structure and function of the transforming genes of human adenoviruses and SV40. Cold Spring Harbor Symp Quant Biol 44:383–399 van der Vliet PC, Levine AJ (1973) DNA-binding proteins specific for cells infected by adenovirus. Nature 246:170–174

van der Vliet PC, Sussenbach JS (1975) An adenovirus type 5 gene function required for initiation of viral DNA replication. Virology 67:415–427

van der Vliet PC, Levine AJ, Ensinger MS, Ginsberg HS (1975) Thermolabile DNA binding proteins from cells infected with a temperature-sensitive mutant of adenovirus defective in viral DNA synthesis. J Virol 15:348–354

Vardimon L, Renz D, Doerfler W (1983) Can DNA methylation regulate gene expression? Cancer Res 84:90–102

Wadell G (1979) Classification of human adenoviruses by SDS-polyacrylamide gel electrophoresis of structural polypeptides. Intervirology 11:47–57

Weber J, Mak S (1972) Synthesis of viral components in hybrids of differentially permissive cells infected with adenovirus type 12. Exp Cell Res 74:423–429

Wilson M, Fraser N, Darnell J (1979) Mapping of RNA initiation sites by high dose of UV irradiation. Evidence for three independent promoters within the left 11% of the Ad2-genome. Virology 94:175–184

Subject Index

The Molecular Biology of Adenoviruses 1

30 Years of Adenovirus Research 1953–1983

Editor: **W. Doerfler**

1983. 69 figures. XII, 232 pages
(Current Topics in Microbiology and Immunology,
Volume 109)
ISBN 3-540-13034-9

Contents:

Springer-Verlag
Berlin
Heidelberg
New York
Tokyo

The Molecular Biology of Adenoviruses 2

30 Years of Adenovirus Research 1953-1983

Editor: **W. Doerfler**

1984. 53 figures. Approx. 290 pages.
(Current Topics in Microbiology and Immunology, Volume 110)
ISBN 3-540-13127-2

Contents:
A. M. Lewis, Jr., J. L. Cook: The Interface between Adenovirus-Transformed Cells and Cellular Immune Response in the Challenged Host. – *A. van der Eb, R. Bernards:* Transformation and Oncogenicity by Adenoviruses. – *K. Fujinaga, K. Yoshida, T. Yamashita, Y. Shimizu:* Organization, Integration, and Transcription of Transforming Genes of Oncogenic Human Adenovirus Types 12 and 7. – *H. van Ormondt, F. Galibert:* Nucleotide Sequences of Adenovirus DNAs. – *A. J. Levine:* The Adenovirus Early Proteins. – T. I. Tikchonenko: Molecular Biology of S16 (SA7) and Some Other Simian Adenoviruses. – *G. Wadell:* Molecular Epidemiology of Adenoviruses. – *B. R. Freifeld, J. H. Lichy, J. Field, R. M. Gronostajski, R. A. Guggenheimer, M. D. Krevolin, Kyosuke Nagata, J. Hurwitz, M. S. Horwitz:* The In Vitro Replication of Adenovirus DNA.

Springer-Verlag
Berlin
Heidelberg
NewYork
Tokyo